Studies in Classification, Data Analysis, and Knowledge Organization

T0181327

For further volumes:
http://www.springer.com/series/1564

Wolfgang Gaul • Andreas Geyer-Schulz

Yasumasa Baba • Akinori Okada

Editors

German-Japanese
Interchange of
Data Analysis Results

 Springer

Editors
Wolfgang Gaul
Andreas Geyer-Schulz
Karlsruhe Institute of
    Technology (KIT)
Karlsruhe
Germany

Yasumasa Baba
The Institute of Statistical Mathematics
Tokyo
Japan

Akinori Okada
Graduate School of
    Management and Information
    Systems
Tama University
Tokyo
Japan

ISSN 1431-8814
ISBN 978-3-319-01263-6   ISBN 978-3-319-01264-3 (eBook)
DOI 10.1007/978-3-319-01264-3
Springer Cham Heidelberg New York Dordrecht London

Springer is part of Springer Science+Business Media (www.springer.com)

Preface

Beyond official relationships, that GfKl (Gesellschaft für Klassifikation) and JCS (Japanese Classification Society) as member societies of IFCS (International Federation of Classification Societies) cultivate, the societies have started to intensify the scientific exchange of ideas, views, and knowledge – first on a personal basis and subsequently via the organization of joint meetings in form of conferences and workshops. After the first joint Japanese-German Symposium in Tokyo (2005) and the second German-Japanese Symposium in Berlin (2006) – results of which were published in a special volume of "Studies in Classification, Data Analysis, and Knowledge Organization" under the title "Cooperation in Classification and Data Analysis" – the third meeting took place together with the annual conference of GfKl in Karlsruhe (2010) followed by the fourth Japanese-German Symposium organized in Kyoto (2012). According to the nature of these meetings there are no restrictions concerning research directions presented at these symposia, all new concepts and activities are welcome on which German and Japanese researchers in the field of data analysis (in the broad sense) are currently working.

Given the just described tradition this volume contains substantially extended and further developed material presented at the last symposia with emphasis on results introduced at Kyoto (2012). As an unambiguous assignment of topics addressed in single contributions is sometimes difficult, the peer-reviewed articles of this volume are grouped in a way that the editors thought appropriate. Three parts concerning "Clustering", "Analysis of Data and Models" and "Applications" have been formed. Within the chosen parts the presentations are listed in alphabetical order with respect to the authors' names. At the end of this volume an index is included that, additionally, should help the interested reader.

Some of the papers which were presented at Karlsruhe (2010) and Kyoto (2012) and are not contained in this volume have been or will still be published elsewhere in the scientific literature (see, e.g., the proceedings of the Karlsruhe GfKl (2012) conference under the title "Challenges at the Interface of Data Analysis, Computer Science, and Optimization" in the Springer series "Studies in Classification, Data Analysis, and Knowledge Organization").

This volume – which can serve as another proof that the longstanding and close relationships between the members of GfKl and JCS are growing further – would not have been possible without the personal assistance of colleagues and friends as well as the financial support of organizations which foster scientific research.

From the German side especially DFG (Deutsche ForschungsGemeinschaft) should be mentioned as without her generous support the participation of German researchers at the Japanese meetings would not have been possible. Additionally, KIT (Karlsruhe Institute of Technology), Institut für Entscheidungstheorie und Unternehmensforschung and Lehrstuhl für Informationsdienste & Elektronische Märkte from Fakultät für Wirtschaftswissenschaften at KIT have to be named.

From the Japanese side, the hospitality extended by the Faculty of Culture and Information Science, Doshisha University, is appreciated whose support made it possible to have Kyoto (2012) in the good environment of the old capital Kyoto. Grateful acknowledgment is made to Mathematical Systems Inc., JAPAN, for its kind support.

Finally, we thank the referees for their comments, suggestions and annotated copies w.r.t. the submitted papers and all those who helped to make the symposia at Karlsruhe (2010) and Kyoto (2012) scientifically important and successful meeting points for our colleagues.

For his support in handling the typesetting process with LaTeX we are very grateful to Maximilian Korndörfer. As always we want to mention Springer Verlag, especially Dr. Martina Bihn, for excellent cooperation in publishing this volume.

Karlsruhe, Germany   Wolfgang Gaul
Karlsruhe, Germany   Andreas Geyer-Schulz
Tokyo, Japan   Yasumasa Baba
Tokyo, Japan   Akinori Okada
March 2013

Contents

Part II Analysis of Data and Models

Part III Applications

Contributors

Y. Baba The Institute of Statistical Mathematics, Tokyo, Japan

D. Baier Chair of Marketing and Innovation Management, Brandenburg University of Technology Cottbus, Cottbus, Germany

H.-H. Bock Institute of Statistics, RWTH Aachen University, Aachen, Germany

R. Decker Department of Business Administration and Economics, Bielefeld University, Bielefeld, Germany

M. Ferreira Institute of Materials Engineering, TU Dortmund University, Dortmund, Germany

A. Geyer-Schulz Information Services and Electronic Markets, Karlsruhe Institute of Technology (KIT), Karlsruhe, Germany

K. Hayashi Graduate School of Environmental Science, Okayama University, Okayama, Japan

CREST, Japan Science and Technology Agency, Tokyo, Japan

E. Hüllermeier Department of Mathematics and Computer Science, University of Marburg, Marburg, Germany

F. Ishioka School of Law, Okayama University, Okayama, Japan

CREST, Japan Science and Technology Agency, Tokyo, Japan

H. Kageyama Kansai University, Takatsuki, Osaka, Japan

H.A. Kestler Research Group of Bioinformatics and Systems Biology, Institute of Neural Information Processing, University of Ulm, Ulm, Germany

Y. Komiya Information Initiative Center, Hokkaido University, Sapporo, Japan

K. Kurakawa National Institute of Informatics, Tokyo, Japan

K. Kurihara Graduate School of Environmental Science, Okayama University, Okayama, Japan

CREST, Japan Science and Technology Agency, Tokyo, Japan

L. Lausser Research Group of Bioinformatics and Systems Biology, Institute of Neural Information Processing, University of Ulm, Ulm, Germany

L. Lüpke Department of Business Administration and Economics, Bielefeld University, Bielefeld, Germany

K. Manabe School of Cultural and Creative Studies, Aoyama Gakuin University, Tokyo, Japan

Y. Matsui Graduate School of Information Science and Technology, Hokkaido University, Sapporo, Japan

H. Minami Information Initiative Center, Hokkaido University, Sapporo, Japan

M. Mizuta Information Initiative Center, Hokkaido University, Sapporo, Japan

H.-J. Mucha Weierstrass Institute for Applied Analysis and Stochastics (WIAS), Berlin, Germany

M. Nakai Department of Social Sciences, College of Social Sciences, Ritsumeikan University, Kyoto, Japan

A. Nakayama Tokyo Metropolitan University, Hachioji-shi, Japan

Y. Nishida Graduate School of Human Sciences, Osaka University, Osaka, Japan

S. Nishisato University of Toronto, Toronto, ON, Canada

A. Okada Graduate School of Management and Information Sciences, Tama University, Tama city, Tokyo, Japan

M. Ovelgönne UMIACS, University of Maryland, College Park, MD, USA

N. Raabe Faculty of Statistics, TU Dortmund University, Dortmund, Germany

B. Raman Department of Radiology, Stanford University School of Medicine, Stanford, CA, USA

C. Rautert Institute of Machining Technology, TU Dortmund University, Dortmund, Germany

S. Rendle University of Konstanz, Konstanz, Germany

D. Schindler Department of Business Administration and Economics, Bielefeld University, Bielefeld, Germany

I. Schmitt Institute of Computer Science, Technical University of Cottbus, Cottbus, Germany

T. Shimokawa University of Yamanashi, Kofu, Japan

H. Suito Graduate School of Environmental Science, Okayama University, Okayama, Japan

CREST, Japan Science and Technology Agency, Tokyo, Japan

Y. Sun National Institute of Informatics, Tokyo, Japan

D.Y. Sze Department of Radiology, Stanford University School of Medicine, Stanford, CA, USA

K. Tanioka Graduate school of Culture and Information Science, Doshisha University, Kyoto, Japan

M. Tsuji Kansai University, Takatsuki, Osaka, Japan

H. Tsurumi College of Business Administration, Yokohama National University, Yokohama, Japan

T. Ueda Department of Radiology, St. Luke's International Hospital, Tokyo, Japan

CREST, Japan Science and Technology Agency, Tokyo, Japan

C. Weihs Faculty of Statistics, TU Dortmund University, Dortmund, Germany

H. Yadohisa Department of Culture and Information Science, Doshisha University, Kyoto, Japan

N. Yamashita Japan Society for the Promotion of Science, Tokyo, Japan

Part I
Clustering

Model-Based Clustering Methods for Time Series

Hans-Hermann Bock

Abstract This paper considers the problem of clustering n observed time series $\mathbf{x}_k = \{ x_k(t) \mid t \in \mathscr{T} \}, k = 1, \ldots, n$, with time points t in a suitable time range $\mathscr{T}$, into a suitable number m of clusters $C_1, \ldots, C_m \subset \{1, \ldots, n\}$ each one comprising time series with a 'similar' structure. Classical approaches might typically proceed by first computing a dissimilarity matrix and then applying a traditional, possibly hierarchical clustering method. In contrast, here we will present a brief survey about various approaches that start by defining probabilistic clustering models for the time series, i.e., with class-specific distribution models, and then determine a suitable (hopefully optimum) clustering by statistical tools like maximum likelihood and optimization algorithms. In particular, we will consider models with class-specific Gaussian processes and Markov chains.

1 Introduction

Clustering methods are designed in order to group objects (persons, texts, pictures,...) into classes (clusters) such that each class contains objects which are 'similar' in some sense, while similarities between objects from different classes are typically small. The underlying information about similarities is provided by data (often: data vectors with qualitative or quantitative variables) that are supposed to be known for all objects. This information is included into a statistical clustering criterion or a mathematical algorithm that computes 'automatically' the desired, hopefully optimum or interpretable clusters.

This paper considers the case where we have n objects $k = 1, \ldots, n$ and each object k is characterized by a time series, i.e., a real-valued function $\mathbf{x}_k :=$

H.-H. Bock (✉)
Institute of Statistics, RWTH Aachen University, Aachen, Germany
e-mail: bock@stochastik.rwth-aachen.de

W. Gaul et al. (eds.), *German-Japanese Interchange of Data Analysis Results*,
Studies in Classification, Data Analysis, and Knowledge Organization,
DOI 10.1007/978-3-319-01264-3_1,
© Springer International Publishing Switzerland 2014

3

$\{x_k(t) | t \in \mathcal{T}\}$ with a suitable set of time points $\mathcal{T}$. More specifically, for continuous time with $\mathcal{T} = [0, T]$ we have $\mathbf{x}_k = \{ x_k(t) \mid 0 \leq t \leq T \}$ while for p discrete observation times $t_1 < t_2 < \cdots < t_p$ we write $\mathbf{x}_k = (x_k(t_1), \ldots, x_k(t_p)) =:$ $(x_{k1}, \ldots, x_{kp})$. Typical examples include:

- n cities where temperature, air pressure, and humidity ($p = 3$) are daily recorded.
- n patients where the diagnostic status (with p variables) is observed over time.
- n stocks whose prices are continuously or daily recorded.
- n genes whose expression levels are measured every 2 h.

In these cases a clustering method should provide groups of objects with a 'similar' structure of time series within each group.

In this paper we do not follow the classical way by first determining an $n \times n$ similarity matrix for the objects and then applying a traditional dissimilarity-based clustering method (e.g., average linkage); see, e.g. Kalpakis et al. (2001), Chouakria and Nagabhushan (2007) and Peng and Müller (2008). In contrast we will describe here some clustering approaches that proceed

1. By first defining a probabilistic model for a time series
2. Assuming a class-specific distribution for objects (time series) within the same class
3. And then determine the unknown parameters and the underlying clustering by maximum likelihood approaches.

Other, eventually more comprehensive surveys are given, e.g., by Liao (2005), Frühwirth-Schnatter (2006, 2011), De la Cruz-Mesía et al. (2008), Pamminger and Frühwirth-Schnatter (2010), McNicholas and Murphy (2010) and Delaigle et al. (2012). The following Sect. 2 presents the general approach, Sects. 3–5 describe models with regression structures, Sect. 6 deals with Markov chain models, Sect. 7 considers the continuous time case, and Sect. 8 comments on methods for determining a suitable number of clusters.

2   Model-Based Clustering Approaches

In model-based clustering the *observed* time series $\mathbf{x}_1, \ldots, \mathbf{x}_n$ are considered as realizations of n independent *random* time series (stochastic processes) $\mathbf{X}_1, \ldots, \mathbf{X}_n$ over the appropriate time domain. While we concentrate on the case of p discrete time points $t_1 < t_2 < \cdots < t_p$ with $\mathcal{T} = \{t_1, \ldots, t_p\}$), we will consider the time-continuous case with an interval $\mathcal{T} = [0, T]$ in Sect. 7 below. The distribution of the random processes has the generic form P_ϑ with a suitable parameter vector ϑ (often with a density $f(\mathbf{x}; \vartheta)$). A clustering model results if the set of objects is supposed to be heterogeneous insofar as it is composed by several hidden classes of objects such the distributions or parameters are class-specific and typically different from class to class. At least two basic approaches for modeling this situation are well-known:

(a) The **fixed-partition model** where we assume the existence of a (specified) number m of (unknown) disjoint classes $C_1, \ldots, C_m \subset \{1, \ldots, n\}$ of objects and of m class-specific parameters $\vartheta_1, \ldots, \vartheta_m$ such that $X_k \sim P_{\vartheta_i}$ for all objects $k \in C_i$ $(i = 1, \ldots, m)$.

(b) The **mixture model** where we assume that $X_1, \ldots, X_n$ are all independent with the mixture distribution $X_k \sim P_{\theta, \pi} := \sum_{i=1}^{m} \pi_i P_{\vartheta_i}$ where the (unknown) vector $\pi = (\pi_1, \ldots, \pi_m)$ contains the prior probabilities π_i of the (unknown) classes. Note that the corresponding posterior probabilities $\pi_i(x_k) := \frac{\pi_i f(x_k; \vartheta_i)}{\sum_{j=1}^{m} \pi_j f(x_k; \vartheta_j)}$ define a degree of membership of object k in the class i and define a clustering when applying a maximum a posteriori (MAP) rule.

Under these model assumptions the unknown $\mathscr{C} = (C_1, \ldots, C_m)$, $\theta := (\vartheta_1, \ldots, \vartheta_m)$ and $\pi := (\pi_1, \ldots, \pi_m)$ are estimated by maximizing the likelihood of the observed time series $x_1, \ldots, x_n$, by a generalized k-means algorithm in case (a) or by an EM algorithm in case (b).

3 Regression-Type Models for Normal Distributions

A basic clustering model results if all time series from some class follow, up to some noise, the same class-specific regression model. The most simple approach assumes a p-dimensional normal distribution for the X_k with independent measurements $X_k(t) \sim \mathscr{N}_1(\mu_i(t), \sigma_i^2(t))$ for $t \in \mathscr{T} = \{1, \ldots, p\}$. Then a class C_i is characterized by

- A class-specific regression function $\mu_i(t) = \mu(t; \vartheta_i)$ and
- A class-specific variance function $\sigma_i^2(t) = \sigma^2(t; \zeta_i)$

with (unknown) parameters ϑ_i and ζ_i. Following the approach (a) this amounts to the more or less classical clustering model for $X_k = (X_k(t_1), \ldots, X(t_p))' := (X_{k1}, \ldots, X_{kp})'$:

$$X_k \sim \mathscr{N}_p(\mu_i, \Sigma_i) \qquad \text{for } k \in C_i \qquad (1)$$

with $\mu_i = (\mu(t_1; \vartheta_i), \ldots, \mu(t_p; \vartheta_i))'$ and the $p \times p$ diagonal covariance matrix $\Sigma_i := diag(\sigma^2(t_1; \zeta_i), \ldots, \sigma^2(t_p; \zeta_i))$. Maximizing the likelihood is equivalent to minimizing the least-squares clustering criterion

$$g(\mathscr{C}; \theta, \zeta) := \sum_{i=1}^{m} \sum_{k \in C_i} \left[\sum_{j=1}^{p} \frac{|x_k(t_j) - \mu(t_j; \vartheta_i)|^2}{\sigma^2(t_j; \zeta_i)} \right] \longrightarrow \min_{\mathscr{C}, \theta, \zeta}. \qquad (2)$$

Typical regression functions include (i) a constant value $\mu(t; \vartheta_i) = \vartheta_i$, (ii) a regression line $\mu(t; \vartheta_i) = a_i + b_i t$ with $\vartheta_i = (a_i, b_i)$, (iii) trigonometric functions such as $\mu(t; \vartheta_i) = c_i \cdot sin(d_i + e_i t)$ with $\vartheta_i = (c_i, d_i, e_i)$, and (iv) truncated

Fourier series or wavelet decompositions. Variances might be, e.g., (i) constants $\sigma^2(t; \zeta_i) = \sigma_i^2$ or σ^2 with $\zeta_i = \sigma_i^2 > 0$, (ii) linear in t as $\sigma^2(t; \zeta_i) = \alpha_i + \beta_i \cdot t$ with $\zeta_i = (\alpha_i, \beta_i)$, etc.

4   Class-Specific Recursive and Autoregressive Models

When considering time series we may, in contrast to the situation in Sect. 3, usually expect dependencies among the measured data values at different time points. Insofar a diagonal covariance matrix in (1) might be too simplistic and should be replaced by an arbitrary class-specific and positive definite covariance matrix $\Sigma_i = (\sigma_{irs}) = (cov(X_k(t_r), X_k(t_s))$ for $k \in C_i$. The corresponding modification of the clustering criterion (2) is then obvious.

In the case of equally spaced time points $t = 1, \ldots, p$ the resulting clustering model can be reformulated in terms of a related regressive or autoregressive process: We use the fact that the covariance matrix Σ_i has a singular value decomposition $\Sigma_i = Q_i' \Lambda_i Q_i$ with a diagonal matrix $\Lambda_i = diag(\sigma_{i1}^2, \ldots, \sigma_{ip}^2)$ and an orthogonal lower triangular $p \times p$ matrix

$$Q_i = (q_{i,rs}) = \begin{pmatrix} 1 & 0 & 0 & \cdots 0 \\ q_{i,21} & 1 & 0 & \cdots 0 \\ q_{i,31} & q_{i,32} & 1 & \cdots 0 \\ \vdots & \vdots & \vdots & \vdots \vdots \\ q_{i,p1} & q_{i,p2} & q_{i,p3} & \cdots 1 \end{pmatrix}.$$

Therefore the linearly transformed variable $\mathbf{Z}_k := Q_i(\mathbf{X}_k - \mu_i)$ has distribution $\mathcal{N}_p(0_p; Q_i Q_i' \Lambda_i Q_i Q_i' = \Lambda_i)$ with p independent normal components $\mathcal{N}_1(0, \sigma_{it}^2)$. Writing down the individual rows $t = 1, \ldots, p$ of $\mathbf{X}_k = \mu_i + Q_i' \mathbf{Z}_k$ shows that the model (1) is identical to the following *recursive model* with class-specific coefficients:

$$X_k(t) = X_{kt} = \mu_{it} + \sum_{\tau=1}^{t} (-q_{i,t,t-\tau})(X_k(t - \tau) - \mu_{i,t-\tau}) + \sigma_{it} U_{it} \qquad (3)$$

for $k \in C_i$, with independent errors $U_{it} \sim \mathcal{N}_1(0, 1)$. Note that for the special case where all elements of Q_i below the s-th subdiagonal are 0 the model (3) reduces to the class-specific autoregressive (AR) process:

$$X_k(t) = X_{kt} = \mu_{it} + \sum_{\tau=1}^{s} (-q_{i,t,t-\tau})(X_k(t - \tau) - \mu_{i,t-\tau}) + \sigma_{it} U_{it}. \qquad (4)$$

Quite generally, AR processes provide a major tool when specifying clustering models for time series (typically formulated conditionally on fixed initial values, e.g., $X_k(0) = x_k(0)$). As an example we mention the classical first order AR process

$$X_k(t) = a_i + b_i X_k(t-1) + \qquad \sigma_h U_{it}. \tag{5}$$

Generalizations are provided by higher order models, and also by dynamically varying clustering models where in addition to the recorded value $X_k(t)$ we observe, at each time point t, a secondary, time-varying, q-dimensional covariable vector $\mathbf{y}_{kt} \in \mathbf{R}^q$:

$$X_k(t) = a_i + b_i X_k(t-1) + \mathbf{c}_i' \, \mathbf{y}_{kt} + \sigma_h U_{it} \tag{6}$$

with class-specific regression coefficient vectors $\mathbf{c}_i \in \mathbf{R}^q$ which are to be determined, e.g., iteratively within the generalized k-means algorithm.

5 Clustering Models with Time-Dependent Regression Regimes

Various authors have considered clustering models where, for each class, there are several different, alternative regression functions (describing different behaviour types, 'regimes', 'options') and objects from this class may shift, from time to time, from one regime to another one. We describe here a model proposed by Samé et al. (2011) which is best illustrated by considering n consumers of electricity and their energy consumption $x_k(t_j)$ at discrete time points t_j. We suppose:

1. There are m classes $C_1, \ldots, C_m$ of consumers (eventually with frequencies $\pi_1, \ldots, \pi_m$).
2. For each class C_i, there exist L (e.g.) polynomial class-specific regression functions of degree q ('options')

$$R_{is}(t) = \beta_{is0} + \beta_{is1} \cdot t + \beta_{is2} \cdot t^2 + \cdots + \beta_{isg} \cdot t^g$$
$$= (1, t, t^2, \ldots, t^g)(\beta_{is0}, \beta_{is1}, \ldots, \beta_{isg})' =: T(t)' \beta_{is} \qquad s = 1, \ldots, L.$$

3. At each time point $t = t_j$, the consumption $X_k(t_j)$ of a consumer $k \in C_i$ is given by $R_{is}(t_j)$ with one of the options s (up to a random error):
   $X_k(t_j) = R_{is}(t_j) + \sigma_{is} U_{is}$ with $U_{is} \sim \mathcal{N}(0, 1)$.
4. The option s is randomly selected: if $k \in C_i$ chooses option s at $t = t_j$ we write:

$$W_{kj} = (W_{kj,1}, \ldots, W_{kj,s}, \ldots, W_{kj,L}) = (0, \ldots, 1, \ldots, 0) =: e_s'$$

5. This random selection process is described by a *multinomial distribution*:

$$W_{kj} \sim Pol(1; q_{is}(t_j; \alpha_i), \ldots, q_{iL}(t_j; \alpha_i))$$

with class-specific time-dependent probabilities from the logistic model

$$q_{is}(t_j; \alpha_i) := P(W_{kj} = e'_s | k \in C_i)$$

$$= \frac{\exp\{\alpha_{is,1}t_j + \alpha_{is,0}\}}{\sum_{h=1}^{L} \exp\{\alpha_{ih,1}t_j + \alpha_{ih,0}\}} \qquad s = 1, \dots, L.$$

Assuming further that consumers, daily consumptions, choices of options, random errors are all independent we obtain the following class-specific distribution density for any time series $X_k = (X_k(t_1), \dots, X_k(t_p))$ from class C_i:

$$f_i(x; \alpha_i, \beta_i) := \prod_{j=1}^{p} \sum_{s=1}^{L} q_{is}(t_j; \alpha_i) \cdot \varphi(x(t_j); T(t_j)'\beta_{is}, \sigma_{is}^2) \qquad (7)$$

where $\varphi(x; \mu, \sigma^2)$ is the density of $\mathcal{N}_1(\mu, \sigma^2)$. Following the general scheme from Sect. 2 these distributions can be combined in the form of a fixed-partition clustering model or, alternatively, by a mixture model, with a subsequent k-means or EM algorithm for estimating parameters and the hidden clusters.

Other approaches for dynamic shifting among different options, within the same class, is proposed in Horenko (2010) in the framework of a fuzzy model with constraints that reduce hectic oscillations between different options over time, and in Song et al. (2008) where dynamically varying and suitably modeled posterior probabilities are used.

6   Clustering with Class-Specific Markov Chain Models

A major tool for describing the dependence structure of stochastic processes is provided by Markov models. Such models have been used in the framework of clustering models as well, see, e.g., De la Cruz-Mesía et al. (2008), Pamminger and Frühwirth-Schnatter (2010) and Frühwirth-Schnatter (2011). In the following we concentrate on the case of Markov chains with T equidistant time points $t = 0, 1, \dots, T$ and values (states, categories) from a finite state space (alphabet) $\mathcal{A} = \{1, \dots, c\}$. This model may apply, e.g., to the evolution of a disease (such as HIV) with c states of disease, observed on T time points for n patients $k = 1, \dots, n$ (cohort study). The model assumes that for objects k from class C_i the time series $X_k = \{X_{kt} \mid t = 0, \dots, T\}$ is a homogeneous Markov chain with

- A class-specific initial distribution:
  $$p_a^{(i)} := P(X_{k0} = a) \qquad a \in \mathcal{A}$$
- Class-specific transition probabilities:
  $$p_{ab}^{(i)} := P(X_{k,t+1} = b | X_{kt} = a) \quad a, b \in \mathcal{A}.$$

The likelihood of the observed time series $\mathbf{x}_k$ is then given by $f(\mathbf{x}_k; \vartheta_i) = P_{\vartheta_i}(\mathbf{X}_k = \mathbf{x}_k) = p_{x_{k0}}^{(i)} \cdot \prod_{t=1}^{T} p_{x_{k,t-1},x_{k,t}}^{(i)}$ with the parameter vector $\vartheta_i := (p_1^{(i)}, \ldots, p_c^{(i)}, p_{11}^{(i)}, \ldots, p_{cc}^{(i)})$.

6.1 Fixed-Partition Clustering Model

Therefore, in the framework of the fixed-partition model, the likelihood of all observed time series $\mathbf{x}_1, \ldots, \mathbf{x}_n$ is given by

$$L(\mathscr{C}, \theta) = \prod_{i=1}^{m} \prod_{k \in C_i} [p_{x_{k0}}^{(i)} \cdot \prod_{t=1}^{T} p_{x_{k,t-1},x_{k,t}}^{(i)}] = \prod_{i=1}^{m} \left\{ \prod_{a=1}^{c} [p_{a0}^{(i)}]^{n_a^{(i)}} \right\} \cdot \left\{ \prod_{a=1}^{c} \prod_{b=1}^{c} [p_{ab}^{(i)}]^{n_{ab}^{(i)}} \right\}$$

where

$n_i \;\; := \;\; |C_i|$, the number of objects (samples) in C_i

$n_a^{(i)} \;\; := \;\; |\{ k \in C_i \mid x_{k0} = a \}|$, the number of samples from C_i starting in state a

$n_{ab}^{(i)} \;\; := \;\; \sum_{k \in C_i} n_{ab,k}^{(i)}$, the number of state transitions $a \to b$ in C_i.

This criterion is to be maximized w.r.t. $\mathscr{C}$ and θ by the k-means algorithm. Starting with an initial m-partition $\mathscr{C} = (C_1, \ldots, C_m)$ of $\{1, \ldots, n\}$, a first step A maximizes L w.r.t. the parameter θ and yields:

$\hat{p}_a^{(i)} := n_a^{(i)}/n_i \qquad$ the relative frequency of time series in C_i starting in a at $t = 0$

$\hat{p}_{ab}^{(i)} := n_{ab}^{(i)}/n_{a\bullet}^{(i)} \qquad$ the relative frequency of observed transitions $a \to b$ in C_i with $n_{a\bullet}^{(i)} := \sum_{b=1}^{c} n_{ab}^{(i)}$ the number of occurrences of state a in C_i (for $t \leq T - 1$).

Afterwards step B determines the maximum-likelihood partition $\mathscr{C}^* := (C_1^*, \ldots, C_m^*)$ with classes:

$$C_i^* := \{ k \mid P_{\hat{\vartheta}_i}(\mathbf{X}_k = \mathbf{x}_k) := \hat{p}_{x_{k0}}^{(i)} \cdot \prod_{t=0}^{T-1} \hat{p}_{x_{kt},x_{k,t+1}}^{(i)} = \max_{j=1,\ldots,m} P_{\hat{\vartheta}_j}(\mathbf{X}_k = \mathbf{x}_k) \}$$

Finally the steps A and B are iterated in turn until stationarity is obtained.

6.2 Mixture Approach

Starting from the densities $f(\mathbf{x}_k; \vartheta_i)$, $i = 1, \ldots, m$, the mixture approach (also termed 'dynamic Bayesian model' in this case) leads to the mixture likelihood

$$G(\mathbf{p}, \mathscr{P}, \pi) := \prod_{k=1}^{n} \left[\sum_{i=1}^{m} \pi_i \cdot p_{x_{k0}}^{(i)} \prod_{t=0}^{T-1} p_{x_{kt},x_{k,t+1}}^{(i)} \right] \rightarrow \max_{\pi, \mathscr{P}, \mathbf{p}}$$

that must be maximized, by an adapted EM algorithm, w.r.t.

- The vector of class frequencies $\pi := (\pi_1, \ldots, \pi_m)$
- The vector $\mathbf{p} = (\pi_1^{(1)}, \ldots, \pi_c^{(m)})$ of all cm initial probabilities
  $\pi_a^{(i)} := P_i(X_{k0} = a)$,
- And w.r.t. the system $\mathscr{P} = (\mathbf{P}^{(1)}, \ldots, \mathbf{P}^{(m)})$ of all transition matrices
  $\mathbf{P}^{(i)} = (p_{ab}^{(i)})$.

Detailed investigations and a survey on this approach are provided by Frühwirth-Schnatter (2006, 2011).

7 Model-Based Methods for Time-Continuous Processes

As seen before it is relatively easy, in the case of discrete time or a discrete state space, to model, write down, and maximize the likelihood of observed time series, in dependence of unknown classifications and parameters. This does no longer hold in the case of real-valued, continuous-time processes $\mathbf{x} = \{x_k(t)|0 \leq t \leq T\}$ where we observe, at least theoretically, infinitely many time points simultaneously. Statistical analysis of such processes is often conducted in the framework of 'functional data analysis' (Ramsay and Silverman 2005; Ferraty and Vieu 2010; Chiou and Li 2007). In case of the clustering problem we describe, in the following, (i) methods that use L_2 theory for random functions and corresponding least-squares criteria (Sect. 7.1), (ii) parametric approaches that model random functions by (truncated) series of suitable orthonormal basis functions (Sect. 7.2), and (iii) methods that employ generalized likelihood concepts for stochastic processes (Sect. 7.3).

7.1 Using L_2 Theory and Least-Squares Criterion

A basic approach generalizes the classical least-squares clustering method to the case of processes $\mathbf{x}_k$ from the Hilbert space $L_2[0, T]$. Assuming the fixed-partition-approach with a given number m classes C_i, we may, e.g., characterize each class by a specific L_2 function $\mu(t; \alpha_i)$ (with a parameter α_i) and minimize the criterion

$$G(\mathscr{C}, \alpha, \beta) := \sum_{i=1}^{m} \sum_{k \in C_i} \int_0^T |x_k(t) - \mu(t; \alpha_i)|^2 \, dt$$

w.r.t. $\mathscr{C}$ and $\alpha_1, \ldots, \alpha_m$ in a generalized k-means algorithm. More refined models for the underlying random processes $\mathbf{X}_k$ may consider class-specific covariance functions $Cov(X_k(t), X_k(\tau)) = R(t, \tau; \beta_i)$ as well even if the consideration of the function R introduces many additional parameters for estimation which may endanger the reliability of the numerical clustering results, but, on the other hand, provides the opportunity to specify here a range of more 'parsimonous' covariance models.

7.2   Using Truncated Series Expansions

The 'infinite-dimensional' case can be reduced to a finite-dimensional one by considering a prespecified (!) complete system of orthonormal $L_2[0, T]$ functions $\{\psi_j(\cdot) | j = 0, 1, \ldots\}$ (e.g., trigonometric functions, wavelets,...) and the corresponding series expansion

$$X_k(t) = \sum_{j=0}^{\infty} A_{kj} \psi_j(t) \approx X_k^{(q)}(t) := \sum_{j=0}^{q} A_{kj} \psi_j(t) \qquad (8)$$

with random Fourier coefficients $A_{kj} := \int_0^T \psi_j(t) X_k(t) dt$ for $j = 0, 1, \ldots$. If q is large enough such that $X_k(t)$ is sufficiently well approximated by the truncated series $X_k^{(q)}(t)$ then we may concentrate on the analysis of the (finite dimensional) random vectors $Y_k := (A_{k0}, A_{k1}, \ldots, A_{kq})' \in \mathbf{R}^{q+1}$ of the first $q + 1$ Fourier coefficients and set up suitable clustering models for $Y_1, \ldots, Y_n$ (with 'observed' values $y_1, \ldots, y_n$ obtained by substituting $x_k(t)$ for $X_k(t)$ in the formula for A_{kj}). So this approach leads to classical multivariate clustering models, eventually under normal assumptions.

While this approach started with a prespecified system of orthonormal functions, some authors used a functional principal component analysis (PCA, the Karhunen-Loève (KL) expansion) for the stochastic processes $X_k(\cdot)$ from L_2. For a process $X(t)$ with $\mu(t) := E[X(t)]$ and $R(t, \tau) := Cov(X(t), X(\tau))$ this expansion has the form

$$X(t) = \mu(t) + \sum_{j=1}^{\infty} A_j \psi_j(t)$$

where ψ_j, λ_j are the eigenfunctions and ordered eigenvalues of the covariance operator R of X:

$$\int_0^T R(t, \tau) \, \psi_j(t) dt = \lambda_j \, \psi_j(t) \qquad j = 1, 2, \ldots$$

and the coefficients (KL scores) $A_j := A_j(X) := \int_0^T \psi_j(t)(X(t) - \mu(t))dt$ are uncorrelated with $E[A_j] = 0$ and $Var(A_j) = \lambda_j \downarrow 0$. Similarly as before, the first q coefficients have been used for characterizing classes in a clustering model. In practice, however, it is difficult to estimate the covariance matrix R from clustered data.

7.3 Using an Approximate Density Concept for Stochastic Processes

It is difficult to design, for continuous-time processes $X_1(t), \ldots, X_n(t)$ with $t \in [0, T]$, a joint likelihood under a clustering model. Therefore Delaigle and Hall (2010) and Jacques and Preda (2012) introduced and used a new 'approximate density concept' in clustering. The basic idea is to find an analogue to the relation $f(x) = \lim_{h \to 0} P(|X - x| \le h)/(2h)$ that holds for any univariate random variable X with a smooth density f, for each value $x \in \mathbf{R}^1$. Measuring the deviation between a random process $X \in L_2$ and a given function $x \in L_2$ by

$$||X - x||_2^2 := \int_0^T |X(t) - x(t)|^2 dt = \sum_{j=1}^{\infty} \lambda_j (A_j - a_j)^2$$

(with the KL scores A_j, a_j of X and x) they find, under suitable assumptions, the following 'approximation theorem for the small-ball probability':

$$\log P(||X - x||_2 \le h) = \sum_{j=1}^{q} \log f_{A_j}(a_j(x)) + \xi(h, r(h)) + o(r(h))$$

with:

- f_{A_j} the density of the random KL score A_j (e.g., normal density)
- $a_j(x) := \int_0^T \psi_j(t)x(t)dt$ the KL scores for function x
- a function $r(h)$ such that $r(h) \to \infty$ for $h \downarrow 0$,
- a function ξ that depends on h and the infinite KL eigenvalue sequence,
- but both r and ξ do not depend on x.

So the first term captures essentially the variation of the l.h.s. with x.

In analogy to the one-dimensional case the expression $f_X^{(q)}(x) = \prod_{j=1}^q f_{A_j}(a_j(x))$ is called the 'approximate density of X at x', with $f_{A_j} \sim \mathcal{N}_1(0, \lambda_j)$ for a centered Gaussian process X.

The resulting fixed-partition clustering model then uses the 'approximate density' in analogy to classical cases: Within each class C_i

- We consider the first q class-specific functional principal components

- Obtained from eigenvalues/eigenvectors $\psi_j^{(i)}, \lambda_j^{(i)}$ of the class-specific covariance function.

Then, for $k \in C_i$, the 'approximate density' of X_k at $x \in L_2$ is

$$X_k \sim f^{(q)}(x) := \prod_{j=1}^{q} f_{A_j^{(i)}}(a_j^{(i)}(x))$$

and the 'approximate likelihood' of the observed time series $x_1, \ldots, x_k$ (with classes $C_1, \ldots, C_m$) is given by

$$L(\mathscr{C}) := \prod_{i=1}^{m} \prod_{k \in C_i} \prod_{j=1}^{q} f_{A_j^{(i)}}(a_j^{(i)}(x_k)).$$

In the special case of n Gaussian processes $X_1, \ldots, X_k$ we find that

$$L(\mathscr{C}) := \prod_{i=1}^{m} \prod_{k \in C_i} \prod_{j=1}^{q} \frac{1}{\sqrt{2\pi}\sigma_j^{(i)}} \cdot exp\left(-\frac{1}{2} \frac{a_j^{(i)}(x_k)^2}{\sigma_j^{(i)}}\right)$$

with class-specific variances $\lambda_j^{(i)} = \sigma_j^{(i)^2}$ of the principal scores. Then a modification of the k-means algorithm is used for maximizing $L(\mathscr{C})$. Thereby, the iterative estimation of the intra-class covariances and the update of the class-specific KL expansions (eigenvalues, eigenvectors) may lead to time-consuming numerical algorithms.

8 Determining the Number of Clusters

Apart from the process of (re-)constructing a clustering of objects, clustering approaches have often to tackle with a series of related problems, e.g., the selection of distribution models, the consideration of outliers, the reduction of information, and the estimation of an appropriate (the 'true') number of clusters. In this last section we summarize some methods for dealing with the latter estimation problem.

In the framework of model-based approaches the determination of an appropriate number of clusters can be considered as a model selection problem. General approaches for model selection are described, e.g., in Claeskens and Hjort (2008) or Wasserman (2000). In particular, we mention here the classical Bayesian model selection approach (using the Bayes factor based on a prior distribution for the number of classes), Akaike's information criterion (AIC; Akaike 1974), the Bayesian Information Criterion (BIC; Schwarz 1978), the Deviance Information Criterion (DIC; Spiegelhalter et al. 2002), the Integrated Classification Likelihood

(ICL; Biernacki et al. 2000), and the Focused Information Criterion (FIC; Claeskens and Hjort 2003) that are typically used when clustering classical vector-valued data.

Even if the time series situation is a little bit more complicated, these same approaches have been successfully described and applied, e.g., by Sebastiani et al. (1999, classical Bayes method), Frühwirth-Schnatter (2006, 2011), Ferrazzi et al. (2005, BIC), De la Cruz-Mesía et al. (2008, BIC and Bayes factor), Jacques and Preda (2012, BIC), Banfield and Raftery (1993) and Biernacki and Govaert (1997) who both used an Approximate Weight of Evidence (AWE). Unfortunately, there are no general results on the performance of these methods that apply to all situations, and detailed simulation studies are needed for assessing the value of each method. For example, when clustering Markov chain data, Pamminger and Frühwirth-Schnatter (2010) and Frühwirth-Schnatter (2011) observed that BIC results in too few clusters for small samples, but overfits the true number of classes in large ones (see also Ferrazzi et al. 2005).

References

Akaike H (1974) A new look at the statistical model identification. IEEE Trans Autom Control 19(6):716–723

Banfield JD, Raftery AE (1993) Model-based Gaussian and non-Gaussian clustering. Biometrics 49(3):803–821

Biernacki C, Govaert G (1997) Using the classification likelihood to choose the number of clusters. Comput Sci Stat 29(2):451–457

Biernacki C, Celeux G, Govaert G (2000) Assessing a mixture model for clustering with the integrated completed likelihood. IEEE Trans Pattern Anal Mach Intell 22(7):719–725

Bouveyron C, Jacques J (2011) Model-based clustering of time series in group-specific functional subspaces. Adv Data Anal Classif 5:281–300

Chiou JM, Li PL (2007) Functional clustering and identifying substructures of longitudinal data. J R Stat Soc B (Stat Methodol) 69(4):679–699

Chouakria AD, Nagabhushan PN (2007) Adaptive dissimilarity index for measuring time series proximity. Adv Data Anal Classif 1:5–21

Claeskens G, Hjort NL (2003) "The focused information criterion" (with discussion). J Am Stat Assoc 98:879–899

Claeskens G, Hjort NL (2008) Model selection and model averaging. Cambridge University Press, Cambridge/New York

De la Cruz-Mesía R, Quintana FA, Marshall G (2008) Model-based clustering for longitudinal data. Comput Stat Data Anal 52(3):1441–1457

Delaigle A, Hall P (2010) Defining probability density for a distribution of random functions. Ann Stat 38:1171–1193

Delaigle A, Hall P, Bathia N (2012) Componentwise classification and clustering of functional data. Biometrika 99(2):299–313

Ferraty F, Vieu P (2010) Nonparametric functional data analysis: theory and practice. Springer, New York

Ferrazzi F, Magni P, Bellazzi R (2005) Random walk models for Bayesian clustering of gene expression profiles. Appl Bioinf 4:263–276

Frühwirth-Schnatter S (2006) Finite mixture and Markov switching models. Springer series in statistics. Springer, New York

Frühwirth-Schnatter S (2011) Panel data analysis: a survey on model-based clustering of time series. Adv Data Anal Classif 5(4):251–280

Horenko I (2010) Finite element approach to clustering of multidimensional time series. SIAM J Sci Comput 32(1):62–83

Jacques J, Preda C (2012) Functional data clustering using density approximation. In: Journées de Statistique de la SFdS, Université Libre de Bruxelles, pp 21–25

Kalpakis K, Gada D, Puttagunta V (2001) Distance measures for effective clustering of ARIMA time-series. In: Proceedings IEEE international conference on data mining, San Jose, pp 273–280

Liao TW (2005) Clustering of time series data – a survey. Pattern Recognit 38(11):1857–1874

McNicholas PD, Murphy TB (2010) Model-based clustering of longitudinal data. Can J Stat 38(1):153–168

Pamminger C, Frühwirth-Schnatter S (2010) Model-based clustering of time series. Bayesian Anal 5:345–368

Peng J, Müller HG (2008) Distance-based clustering of sparsely observed stochastic processes, with applications to online auctions. Ann Appl Stat 2:1056–1077

Ramsay J, Silverman BW (2005) Functional data analysis, 2nd edn. Springer series in statistics, Springer, New York

Samé A, Chamroukhi F, Govaert G, Aknin P (2011) Model-based clustering and segmentation of time series with changes in regime. Adv Data Anal Classif 5(4):301–321

Schwarz G (1978) Estimating the dimension of a model. Ann Stat 6(2):461–464

Sebastiani P, Ramoni M, Cohen P, Warwick J, Davis J (1999) Discovering dynamics using Bayesian clustering. In: Hand D, Kok J, Berthold M (eds) Advances in intelligent data analysis. Lecture notes in computer science, vol 1642. Springer, Berlin, pp 199–209

Song X, Jermaine C, Ranka S, Gums J (2008) A Bayesian mixture model with linear regression mixing proportions. In: Proceedings of the 14th ACM SIGKDD international conference on knowledge discovery and data mining, KDD'08, Las Vegas. ACM, New York, pp 659–667

Spiegelhalter DJ, Best NG, Carlin BP, van der Linde A (2002) Bayesian measures of model complexity and fit. J R Stat Soc 64(4):583–639

Vilar JA, Pértega S (2004) Discriminant and cluster analysis for Gaussian stationary processes: local linear fitting approach. J Nonparametr Stat 16:443–462

Wasserman L (2000) Bayesian model selection and model averaging. J Math Psychol 44(1):92–107

The Randomized Greedy Modularity Clustering Algorithm and the Core Groups Graph Clustering Scheme

Andreas Geyer-Schulz and Michael Ovelgönne

Abstract The modularity measure of Newman and Girvan is a popular formal cluster criterium for graph clustering. Although the modularity maximization problem has been shown to be NP-hard, a large number of heuristic modularity maximization algorithms have been developed. In the 10th DIMACS Implementation Challenge of the Center for Discrete Mathematics & Theoretical Computer Science (DIMACS) for graph clustering our core groups graph clustering scheme combined with a randomized greedy modularity clustering algorithm won both modularity optimization challenges: the Modularity (highest modularity) and the Pareto Challenge (tradeoff between modularity and performance). The core groups graph clustering scheme is an ensemble learning clustering method which combines the local solutions of several base algorithms to form a good start solution for the final algorithm. The randomized greedy modularity algorithm is a non-deterministic agglomerative hierarchical clustering approach which finds locally optimal solutions. In this contribution we analyze the similarity of the randomized greedy modularity algorithm with incomplete solvers for the satisfiability problem and we establish an analogy between the cluster core group heuristic used in core groups graph clustering and a sampling of restart points on the Morse graph of a continuous optimization problem with the same local optima.

A. Geyer-Schulz (✉)
Information Services and Electronic Markets, Karlsruhe Institute of Technology (KIT), Karlsruhe, Germany
e-mail: andreas.geyer-schulz@kit.edu

M. Ovelgönne
UMIACS, University of Maryland, College Park, MD, USA
e-mail: mov@umiacs.umd.edu

W. Gaul et al. (eds.), *German-Japanese Interchange of Data Analysis Results*,
Studies in Classification, Data Analysis, and Knowledge Organization,
DOI 10.1007/978-3-319-01264-3_2,
© Springer International Publishing Switzerland 2014

1 Introduction

Newman and Girvan (2004) introduced modularity clustering as an efficient way to find communities with maximal modularity in a large network. Formally, the network is defined by a simple, undirected graph $G = (V, E)$ with the set of vertices V and the set of edges $E \subseteq V \times V$. Alternatively, a simple, undirected graph can be given by its $| V | \times | V |$ adjacency matrix M which is symmetric ($m_{kl} = m_{lk}$), binary, and $m_{kk} = 0$. For an undirected edge $\{v_x, v_y\}$ between the vertices $v_x \in V$ and $v_y \in V$, 2 elements of the adjacency matrix are set to 1, namely $m_{xy} = m_{yx} = 1$.

The communities form a partition $C = \{C_1, \ldots, C_p\}$ of V with p subsets (clusters, communities, groups). We denote the set of all partitions by Ω, for partitions we use upper indices, e.g. C^1, C^{new}, C^{old}, and to refer to the j-th cluster of partition C^{old} we use C_j^{old}. In addition, it is convenient to refer to the cluster to which a vertex v belongs in a partition C as $c_C(v)$.

For a simple, undirected graph G and a partition C with a given number of groups p, the modularity measure $Q(G, C)$ is defined as:

$$Q(G, C) = \sum_{i=1}^{p} (e_{ii} - a_i^2) \tag{1}$$

with

$$e_{ij} = \frac{\sum_{v_x \in C_i} \sum_{v_y \in C_j} m_{xy}}{2 \mid E \mid} \tag{2}$$

and

$$a_i = \sum_j e_{ij}. \tag{3}$$

e is the $p \times p$ weight matrix of partition C: e_{ij} is the fraction of edges between clusters C_i and C_j. e_{ii} is the fraction of edges in cluster C_i. For the singleton partition (each vertex is a cluster), we have $e_{ii} = 0$.

a_i is the fraction of edges that link to vertices in cluster C_i. The fraction of edges with both vertices in cluster C_i when edges are randomly established between vertices is a_i^2. Thus, modularity is the sum over all communities of the difference between the fraction of edges in the same community (e_{ii}) minus the expected value of the fraction of edges (a_i^2) of a network with the same community partition but randomly generated edges. Modularity measures the non-randomness of a graph partition.

The modularity maximization problem is then:

$$\max_{C \in \Omega} Q(G, C) \tag{4}$$

Integer linear programming algorithms solve the modularity maximization problem for small graphs (see e.g. Agarwal and Kempe 2008; Brandes et al. 2007). Brandes et al. (2008) have given an integer linear programming formulation for modularity clustering and established that the formal problem is – in the worst case – NP-hard. However, Smale's analysis of the average case shows that the number of pivots required to solve a linear programming problem grows in proportion to the number of variables on the average (Smale 1983). From the empirical results shown in Sect. 2, we conjecture that Smale's results hold for modularity optimization too.

The fastest heuristic algorithms for modularity maximization so far are greedy agglomerative hierarchical clustering algorithms (see e.g. Schuetz and Caflisch 2008; Zhu et al. 2008). The recent success of the randomized greedy (RG) algorithm with the core groups graph clustering (CGGC) scheme as ensemble learning variant (see Ovelgönne and Geyer-Schulz 2013, 2012b; Ovelgönne et al. 2010) in the 10th DIMACS Implementation Challenge shows that this algorithm currently is the best heuristic algorithm available with regard to speed and nearness to optimality. The question is: Why?

The rest of this paper is an attempt to answer this question: In Sect. 2 we first introduce the RG algorithm with the CGGC scheme and we present some of the performance results for this algorithm.

The RG algorithm with the CGGC scheme actually combines two different ideas:

1. The RG algorithm is a non-deterministic highly efficient gradient algorithm. We relate the analysis of this algorithm to the analysis of randomized solvers (Biere et al. 2009) of the generalized satisfiability (GSAT) problem in Sect. 3.
2. The CGGC-scheme combines k local optima (or almost local optima) to find new start points for local optimization algorithms. We will use the basic idea of Morse theory to explain that the CGGC-scheme heuristically selects points on or near the Morse graph as restart points in Sect. 4. For an introduction to global analysis and Morse theory, see Jongen et al. (2000).

2 Randomized Greedy Modularity Optimization with Core Groups

2.1 The RG Algorithm

Figure 1 shows the pseudocode of the main subroutine of the RG algorithm. The RG algorithm takes a graph G and a partition C as arguments. In this subsection, we start the RG algorithm from a singleton partition: $RG(G, C^{singleton})$. The RG algorithm consists of four phases, namely the initialization of e and a for the partition C (line 1), building of a dendrogram (line 2), extracting the partition

RG(G, C)
Input: Undirected, connected graph G, partition C.
Output: Partition P.
Local: Join list JL, matrix of fractions e, vector a of rowsums of e, modularity Q, maximal
modularity $maxQ$ at level $optLev$.
    [1] $\{_1$ $(e, a, Q) \leftarrow$ RG_Initialize(G, C);
    [2] $(JL, optLev, maxQ) \leftarrow$ RG_BuildDendrogram(G, e, a, Q);
    [3] $P \leftarrow$ RG_ExtractClusters$(G, JL, optLev)$;
    [4] $P \leftarrow$ RG_Refine(P);
    [5]   **return** P;$\}_1$

Fig. 1 Algorithm 1: RG – main subroutine

RG_Initialize(G, C)
Input: Undirected, connected graph $G(V, E)$, partition C.
Output: Matrix of fractions e, vector a of rowsums of e, modularity Q.
Local: Vertices v, w; clusters C_c, C_n; indices c, n.
Functions: sum(vector), rowsum(matrix, index) , diag(matrix).
    [01] $\{_1$ **if** $(C == C^{singleton})$ $\{_2$ **forall** $(v \in V)$
    [02] $\{_3$ **forall** $(neighbors\ w\ of\ v)$ $\{_4$ $e[v, w] \leftarrow 1/(2* |E|)$;$\}_4$
    [03] $a[v] \leftarrow rowsum(e, v)$;$\}_3$
    [04] $Q \leftarrow -sum(a^2)$; $\}_2$
    [05]   **else** $\{_5$ **forall**(clusters $C_c \in C$)
    [06] $\{_6$ **forall** (neighboring clusters C_n of C_c)
    [07] $\{_7$ $x \leftarrow$ number of edges connecting vertices in C_c with vertices in C_n;
    [08] $e[c, n] \leftarrow x/(2* |E|)$; $\}_7$
    [09] $e[c, c] \leftarrow$ (number of edges in cluster C_c)/ $|E|$;
    [10] $a[c] \leftarrow rowsum(e, c)$;$\}_6$
    [11] $Q = sum(diag(e) - a^2)$; $\}_5$
    [12]   **return** (e, a, Q); $\}_1$

Fig. 2 Algorithm 2: RG_Initialize

with the highest modularity (line 3) and searching for a refinement by vertex swaps
(line 4). The RG algorithm returns the best partition found (line 5).

Phase 1. The pseudocode of the subroutine RG_Initialize(G,C) for the initialization
of the weight matrix e and the vector of the rowsums a for a partition C of the
vertices of the graph G is shown in Fig. 2. Functions used are: sum(v) returns the
sum of the elements of the vector v, rowsum(e, i) returns the sum of the i-th row of
matrix e, and diag(e) returns the vector of the main diagonal of the matrix e.

1. Lines 1–3 build up e_{ij} (definition 2) and a_i (definition 3) for the partition $C^{singleton}$
   of singletons. In line 4, the modularity Q of the singleton partition is computed.
2. Lines 5–10 compute e_{ij} and a_i when the algorithm starts with an arbitrary, non-
   singleton partition. For arbitrary partitions, e_{ij}, $i \neq j$ are the fractions of edges
   between clusters (computed in lines 5–8). $e_{ij} > 0$, if clusters c_i and c_j are
   neighboring clusters. Neighboring clusters are clusters for which at least one
   edge with one vertex in c_i and and the second vertex in c_j exists. e_{ii} is the fraction
   of edges in cluster i (computed in line 9). In line 11, we compute the modularity
   Q of the partition. This part of the code is used in line 3 of the CGGC algorithm
   shown in Fig. 7.

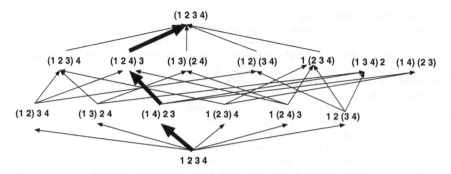

Fig. 3 Join path of RG_BuildDendrogram

Phase 2. The subroutine RG_BuildDendrogram is a randomized hierarchical agglomerative modularity clustering algorithm which when called with a singleton partition $C^{singleton}$ as second argument starts with the n vertices of the graph as singleton clusters ($p = n$) and applies the join operation until a single cluster ($p = 1$) is reached. It constructs a complete dendrogram represented as a list of joins. A sample join path of this algorithm is shown in Fig. 3. Repeated execution of RG_BuildDendrogram produces different join paths.

The pseudocode of RG_BuildDendrogram is shown in Fig. 4. Functions used are: random(s) returns a random element of the set s, without(s, e) returns the set s without the element e, length(v) returns the length of the vector v, append(v, e) appends the element e to the end of vector v.

The subroutine join(i, j) performs the join operation of two clusters in the form of an inplace update of e: $e[i;] = e[i;] + e[j;]$ (sum of rows i and j) and $e[;i] = e[;i] + e[;j]$ (sum of columns i and j) and a: $a[i] = a[i] + a[j]$ (line 7 of RG_BuildDendrogram). For efficiency reasons, this is implemented by a sparse matrix package.

At the heart of RG_BuildDendrogram are two facts which are both exploited in line 5:

1. Merging two unconnected clusters always results in a negative modularity change. Thus the search is restricted to connected clusters.
2. Merging two clusters C_i and C_j changes the modularity of the clustering by $\Delta Q(i, j) = e_{ij} + e_{ji} - 2a_i a_j = 2(e_{ij} - a_i a_j)$. This allows an efficient computation of modularity.

However, the key innovation of this algorithm is the random selection of a cluster combined with a limited search instead of the search for the steepest gradient (line 4). Because only one or two communities are randomly selected (line 3), the effects of this innovation are a dramatic gain in speed and a non-deterministic behavior of the algorithm. Non-deterministic behavior means that the algorithm will move to one of several different local optima (if there are more than one), but it cannot be said from the outset to which one. RG_BuildDendrogram is not locally

RG_BuildDendrogram(G, e, a, Q)
Input: Graph G, matrix of fractions e, vector a of rowsums of e, modularity Q.
Output: vector of join pairs JL, height of dendrogram $optLev$ with maximal modularity $maxQ$.
Local: Number of clusters searched k; indices of clusters $c1$, $c2$; modularity change ΔQ, $max\Delta Q$;
    join pairs $next\,join$; set of active indices $active$.
Functions: random(set), length(vector), append(vector, element), without(set, element).
Subroutines: join(join pair).

```
[01]    {₁ optLev ← 0; maxQ ← Q; active ← {1,...,length(a)};
[02]    for (i = 1 to rank(e) − 1) {₂ maxΔQ ← −∞;
[03]        if (i < rank(e)/2) k ← 1 else k ← 2;
[04]        for (j = 1 to k) {₃ c1 ← random(active);
[05]            forall (clusters c2 linked to c1) {₄ ΔQ ← 2(e[c1,c2] − (a[c1] ∗ a[c2]));
[06]                if (ΔQ > maxΔQ) {₅ maxΔQ ← ΔQ; next join ← (c1,c2);}₅}₄}₃
[07]        join(next join); active ← without(active,c2);
[08]        JL ← append(JL,next join); Q ← Q + maxΔQ;
[09]        if (Q > maxQ) {₆ maxQ ← Q; optLev ← i; }₆}₂
[10]    return (JL, optLev, maxQ);}₁
```

Fig. 4 Algorithm 3: RG_BuildDendrogram

RG_ExtractClusters($G, JL, optLev$)
Input: Graph $G = (V, E)$, vector join pairs JL, height of dendrogram $optLev$ with maximal modularity.
Output: Partition P.
Local: vertices v, w, vector of sets $clusters$, set c.

```
[1]    {₁ forall v ∈ V {₂ clusters[v] ← {v} }₂
[2]    for (i = 1 to optLev) {₂
[2]        (v,w) ← JL[i] {₂
[3]        clusters[v] ← clusters[v] ∪ clusters[w];
[4]        clusters[w] ← ∅;}₂
[5]    P ← {c ∈ clusters|c ≠ ∅};
[6]    return P;}₁
```

Fig. 5 Algorithm 4: RG_ExtractClusters

join-optimal, because as implemented (lines 2 and 3 of RG_BuilDendrogram) the neighbors of at most 1 or 2 clusters are explored.

The algorithm always executes the locally best join, even if the best join has a negative ΔQ and thus leads to a decrease in modularity. However, changes in modularity along a join path are highly irregular: In Fig. 12 only a single join (1, 2) with an increase in modularity exists for the first join, some points (the partition 1 (2 3 4) with the second highest modularity) are not reachable on a join path with increasing modularity, and, last but not least, the optimum can also be reached by following a join path which starts with a decrease in modularity.

Phase 3. The extraction of the partition with the highest modularity $Q(G, C)$ from the dendrogram is implemented by the subroutine RG_ExtractClusters(joinlist). The pseudocode for this subroutine is shown in Fig. 5.

RG_Refine(*C*)
Input: Undirected, connected graph *G*, partition *C*.
Output: Partition *C*.
Local: Boolean flag *change*; modularity change ΔQ, $max\Delta Q$; clusters C_c, C_n, C_*; vertex v.
Function: moveΔQ(vertex, from_cluster, to_cluster).
Subroutine: move(vertex, from_cluster, to_cluster).

```
[1]   {₁ change ← true;
[2]   while (change) {₂ change ← false;
[3]     forall (v ∈ V) {₃
[4]       Cc ← currentCluster(v);
[5]       maxΔQ ← 0;
[6]       forall (neighboring clusters Cn of v) {₄
[7]         ΔQ ← moveΔQ(v,Cc,Cn);
[8]         if (ΔQ > maxΔQ) {₅ maxΔQ ← ΔQ; C* ← Cn; }₅}₄
[9]       if (maxΔQ > 0) {₆ move(v,Cc,C*); change ← true; }₆}₃}₂
[10]  return (C); }₁
```

Fig. 6 Algorithm 5: RG_Refine

Phase 4. The local refinement of this partition by greedy single vertex moves between clusters is implemented by the subroutine RG_Refine whose pseudocode is shown in Fig. 6. In the inner loop (lines 6–8) this algorithm searches for the best move of vertex v to a neighboring cluster. The change in modularity when moving vertex v from one cluster to the other cluster is computed by the function moveΔQ (line 7). If the best move of vertex v leads to an improvement of the modularity measure, the move is executed by the function move (line 9). Because the search for the best change in a neighboring cluster is continued until no further improvement is possible, we call RG_Refine locally 1-change optimal. And the RG-algorithm is also locally 1-change optimal.

2.2 Core Groups and Core Group Partitions

The second idea of combining locally optimal or almost optimal partitions to find new start points for the algorithm is implemented by the $CGGC_{RG}$-algorithm (see Fig. 7).

The $CGGC_{RG}$ algorithm first uses the RG-algorithm (see Fig. 1) to compute a vector of z locally (1-change) optimal solutions (line 1). The subroutine getCoreGroups (line 2) combines the partitions $C^1, \dots, C^z$ into a new partition C^{core} (called core group partition) in the following way:

$$\forall v, w \in V : (\bigwedge_{i=1}^{z} c_{C^i}(v) = c_{C^i}(w)) \Leftrightarrow c_{C^{core}}(v) = c_{C^{core}}(w) \tag{5}$$

$CGGC_{RG}(G)$
Input: Graph G, number of partitions for core group generation z.
Output: Partition C^{best}.
Local: Vector of z partitions C, core group partition C^{core}, singleton partition $C^{singleton}$.
[1] $\{_1$ **for** $(i = 1$ **to** $z)$ $\{_2$ $C[i] \leftarrow$ RG$(G, C^{singleton})$;$\}_2$
[2] $C^{core} \leftarrow$ getCoreGroups(C);
[3] $C^{best} \leftarrow$ RG(G, C^{core}); $\}_1$

Fig. 7 Algorithm 6: $CGGC_{RG}$

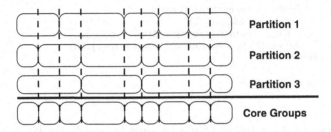

Fig. 8 Example how core groups are extracted from several clusterings

A cluster of a core group partition is called a core group, because all members of a core group are grouped together in the same cluster in each of the z locally optimal partitions from which the core group partition is generated (Eq. 5).

The partition C^{core} is the maximum overlap of a set (an ensemble) of partitions and it serves as a restart point for the randomized greedy algorithm (line 3 of $CGGC_{RG}$). Figure 8 illustrates this operation for three partitions. However, in Ovelgönne and Geyer-Schulz (2012b), $CGGC_{RG}$ shown in Fig. 7 is generalized to an ensemble learning scheme working with arbitrary weak learning algorithms and with an iteration of the core group generation phase. The performance of this algorithm crucially depends on z, the number of partitions used in building core groups. In the 10th DIMACS implementation challenge, this parameter was set to $z = \ln |V|$ based on a preliminary analysis of the dependence of the algorithm's performance on this parameter for a small sample of data sets.

Table 1 shows references w.r.t. test data sets used in the experiments whose results are shown in Table 2. The CGGC scheme and its iterated application combined with the RG algorithm as base and final learner (denoted as $CGGC_{RG}$ respectively $CGGCi_{RG}$) lead to a further improvement in solution quality. For comparison purposes, results of the variable neighborhood search (VNS) algorithm (Aloise et al. 2012) have been included. This algorithm achieved the 2nd best results (after the $CGGCi_{RG}$ algorithm) in the modularity optimization challenge of the 10th DIMACS Implementation Challenge. For small networks, an algorithm with extensive local search like VNS can find competitive or even better results than the $CGGCi_{RG}$ algorithm. The main reason for this is that RG_BuildDendrogram is not locally join-optimal. However, for larger networks the local search approach provides inferior results.

Table 1 Test datasets. A selection of networks from the testbed of the 10th DIMACS implementation challenge available at http://www.cc.gatech.edu/dimacs10/downloads.shtml

Name	Vertices	Edges	Reference
adjnoun	112	425	Newman (2006)
celegans_metabolic	453	2,025	Duch and Arenas (2005)
Email	1,133	10,902	Guimerà et al. (2003)
polblogs	1,490	16,715	Adamic and Glance (2005)
netscience	1,589	2,742	Newman (2006)
power	4,941	6,594	Watts and Strogatz (1998)
hep-th	8,361	15,751	Newman (2001)
PGPgiantcompo	10,680	24,316	Boguñá et al. (2004)
astro-ph	16,706	121,251	Newman (2001)
cond-mat	16,726	47,594	Newman (2001)
coAuthorsCiteseer	22,732	814,134	Geisberger et al. (2008)
cond-mat-2003	31,163	120,029	Duch and Arenas (2005)
citationCiteseer	268,495	1,156,647	Geisberger et al. (2008)
coAuthorsDBLP	299,067	977,676	Geisberger et al. (2008)
coPapersCiteseer	434,102	16,036,720	Geisberger et al. (2008)
coPapersDBLP	540,486	15,245,729	Geisberger et al. (2008)
eu-2005	862,664	16,138,468	Boldi et al. (2011)
in-2004	1,382,908	13,591,473	Boldi et al. (2011)

Table 2 Comparison of the average and best modularity identified by the algorithms RG, $CGGC_{RG}$, $CGGCi_{RG}$ and VNS. Results are compiled from Ovelgönne and Geyer-Schulz (2012b) and Aloise et al. (2012). The best value is typeset in bold

	Average modularity obtained				Best modularity obtained			
	RG	$CGGC_{RG}$	$CGGCi_{RG}$	VNS	RG	$CGGC_{RG}$	$CGGCi_{RG}$	VNS
adjnoun	0.2925	0.3061	0.3062	**0.3134**	0.3072	0.3111	0.3119	**0.3134**
celegans_metabolic	0.4367	0.4502	0.4502	**0.4532**	0.4485	0.4528	0.4523	**0.4532**
Email	0.5712	0.5799	0.5801	**0.5826**	0.5787	0.5821	0.5826	**0.5828**
polblogs	0.4258	0.4268	0.4268	**0.4271**	0.4260	**0.4271**	**0.4271**	**0.4271**
netscience	0.9404	0.9598	0.9597	**0.9599**	0.9499	**0.9599**	**0.9599**	**0.9599**
power	0.9282	0.9396	0.9397	**0.9408**	0.9333	0.9404	0.9404	**0.9409**
hep-th	0.8340	0.8558	0.8557	**0.8576**	0.8425	0.8565	0.8566	**0.8577**
PGPgiantcompo	0.8644	**0.8862**	**0.8862**	0.8860	0.8746	**0.8866**	0.8865	0.8860
astro-ph	0.6970	0.7428	0.7428	**0.7446**	0.7148	0.7442	0.7444	**0.7449**
cond-mat	0.8298	0.8524	0.8524	**0.8532**	0.8368	0.8531	0.8530	**0.8534**
coAuthorsCiteseer	0.8951	**0.9051**	**0.9051**	0.8330	0.8963	**0.9053**	0.9052	0.8336
cond-mat-2003	0.7571	0.7775	**0.7776**	0.7764	0.7618	**0.7786**	**0.7786**	0.7767
citationCiteseer	0.8086	0.8233	**0.8234**	0.8203	0.8119	0.8239	**0.8241**	0.8207
coAuthorsDBLP	0.8208	0.8373	**0.8406**	0.8330	0.8222	0.8382	**0.8411**	0.8336
coPapersCiteseer	0.9163	0.9217	**0.9222**	0.9204	0.9175	0.9221	**0.9225**	0.9205
coPapersDBLP	0.8538	0.8647	**0.8666**	0.8610	0.8556	0.8653	**0.8668**	0.8615
eu-2005	0.9390	**0.9411**	**0.9411**	**0.9411**	0.9403	**0.9416**	**0.9416**	0.9414
in-2004	0.9776	0.9783	**0.9806**	0.9805	0.9785	0.9795	**0.9806**	0.9805

GSAT(*F*, *MAX_FLIPS*, *MAX_TRIES*)
Input: CNF formula *F*; Parameters: *MAX_FLIPS*, *MAX_TRIES*: Integer
Output: A satisfying assignment ρ for *F* or *FAIL*
[1]   {$_1$ **for** ($i = 1$ **to** *MAX_TRIES*) {$_2$
[2] $\rho \leftarrow$ a randomly assigned truth assignment for *F*;
[3]   **for** ($j = 1$ **to** *MAX_FLIPS*) {$_3$
[4]   **if** (ρ satisfies *F*) {$_4$ **return**(ρ);}$_4$
[5] $v \leftarrow$ a variable flipping which results in the greatest decrease (possibly negative)
[6]   in the number of unsatisfied clauses;
[7]   Flip v in ρ; }$_3$}$_2$
[8]   **return**(*FAIL*); }$_1$

Fig. 9 Algorithm 7: GSAT

3 Randomized SAT Solvers and the RG Algorithm: Local Search

In this section we introduce the GSAT and the WALKSAT algorithms as the most prominent members of the family of incomplete randomized SAT solvers and compare these algorithms with the RG algorithm. We introduce discrete Lagrangian methods, because they offer a common formal framework for the analysis of incomplete SAT solvers. In the rest of this section we will borrow parts of this framework for the analysis of the RG algorithm.

The satisfiability problem is formulated as follows: *F* is an *n*-variable conjunctive normal form (CNF) formula with clauses $C_1, C_2, \ldots, C_m$. For the satisfiability problem of the propositional formula *F*, the discrete manifold (its solution landscape) is defined as $\{0, 1\}^n \times \{0, 1, \ldots, m\}$ where the first term in the Cartesian product denotes a truth assignment (the point *x* in $\{0, 1\}^n$) to the *n* variables and the second term $\{0, 1, \ldots, m\}$ the number of clause violations of the point *x* for *F*. A truth assignment with zero violated clauses is a global minimum in the discrete manifold and a solution of the satisfiability problem.

Both GSAT and WALKSAT algorithms have the property that despite the NP-completeness of the GSAT problem, they often solve GSAT problems very fast. This discovery was published as early as 1979 by Goldberg and this property of the satisfiability problem led to a massive research effort in theoretical computer science, discrete mathematics, and theoretical physics (Biere et al. 2009). For a survey on incomplete SAT-solvers like the GSAT and WALKSAT algorithms, see Kautz et al. (2009).

Figure 9 shows the pseudocode of the GSAT algorithm. The GSAT algorithm starts its search from a random initial truth-assignment (line 2) by flipping the truth value of the variable which leads to the largest reduction in the number of clause violations (inner loop, lines 3–7). The GSAT algorithm is a discrete deterministic gradient method with repeated runs from a random restart point (outer loop, lines 1–7). The GSAT algorithm can solve GSAT problems which are one order of magnitude larger than those solved by complete algorithms (e.g. resolution

WALKSAT (*F*, *MAX_FLIPS*, *MAX_TRIES*, *p*)
Input: CNF formula *F*; Parameters: *MAX_FLIPS*, *MAX_TRIES*: Integer; noise $p \in [0,1]$
Output: A satisfying assignment ρ for *F* or *FAIL*
```
[1]   {₁for (i = 1 to MAX_TRIES) {₂
[2]     ρ ← a randomly assigned truth assignment for F;
[3]     for (j = 1 to MAX_FLIPS) {₃
[4]       if (ρ satisfies F) {₄ return(ρ) }₄
[5]       C ← an unsatisfied clause of F chosen at random;
[6]       if (∃ variable x ∈ C with break_count = 0) {₅ Free move: v ← x; }₅
[7]       else {₆ Metropolis move: With p: v ← a variable ∈ C chosen at random;
[8]         Greedy move: With 1 − p: v ← a variable ∈ C with the smallest break_count;}₆
[9]       Flip v in ρ; }₃}₂
[10]  return(FAIL); }₁
```

Fig. 10 Algorithm 8: WALKSAT

or backtrack algorithms). The main disadvantage of GSAT is its inability to detect infeasible problems.

The WALKSAT algorithm whose pseudocode is shown in Fig. 10 interleaves the greedy moves of GSAT – flipping a variable in the clause which minimizes the number of currently satisfied clauses that become unsatisfied (the break-count) – with random walk moves from a Metropolis search, it focuses its search by always selecting the variable to flip from an unsatisfied clause C chosen at random. The Metropolis moves of the WALKSAT algorithm imply that the WALKSAT algorithm can – with a positive probability – pass through valleys in the search space and, when combined with an optimal cooling schedule (Hajek 1988), find a global optimum asymptotically. The WALKSAT algorithm is similar to the RG algorithm in its random selection of a local neighborhood (the variables in an unsatisfied clause).

Next, we present discrete Lagrangian methods for the SAT problem. For x as defined above, $U_i(x)$ is a function that is 0 if C_i is satisfied, and 1 otherwise. $U(x)$ denotes the m-dimensional vector of clause violations. The SAT problem in its constrained formulation is $\min_x N(x) = \sum_{i=1}^{m} U_i(x)$ subject to $U_i(x) = 0$ $\forall i \in \{1, 2, \ldots, m\}$. We obtain an unconstrained optimization problem by the discrete Lagrangian function (Shang and Wah 1998): $L_d(x, \lambda) = N(x) + \sum_{i=1}^{m} \lambda_i U_i(x)$ with $\lambda = (\lambda_1, \lambda_2, \ldots \lambda_m) \in R^m$. The first term of the Lagrangian function is the total number of constraint violations, the second term the sum of the weighted violations in non-satisfied clauses. The redundancy of this formulation is its main strength, because the Lagrangian multipliers λ_i (one for each constraint) introduce a clause weighting scheme by introducing penalties to violated clauses. For the SAT problem, all local minima are also global minima and, therefore, the discrete Lagrangian methods can find the global optimum. However, for modularity maximization this property does not hold.

Discrete Lagrangian algorithms for the SAT problem rely on finding a saddle point. A point $(x^*, \lambda^*) \in \{0, 1\}^n \times R^m$ is a saddle point of $L_d(x, \lambda)$ if it is a local minimum with regard to x^* and a local maximum with regard to λ^*. Formally, (x^*, λ^*) is a saddle point for $L_d(x, \lambda)$ if $L_d(x^*, \lambda) \leq L_d(x^*, \lambda^*) \leq L_d(x, \lambda^*)$ for

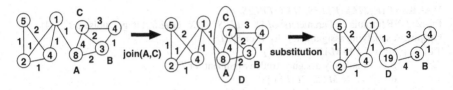

Fig. 11 Joins as graph rewriting

λ sufficiently close to λ^* and for all x that differ from x^* in only one dimension. x^* is a local minimum for the constrained SAT problem defined above, if there exists a λ^* so that (x^*, λ^*) is a saddle point of $L_d(x, \lambda)$. For a proof, see Shang and Wah (1998, pp. 69–70).

A gradient algorithm for the SAT problem finds such saddle points by doing descents in x and ascents in λ. However, to implement such an algorithm, we have to construct a discrete difference gradient: The neighborhood $N(x)$ of x contains x and all points y one variable flip away: $y = x + v$ with v the direction vector in $\{-1, 0, 1\}^n$ with at most one non-zero entry $v_i = \begin{cases} 1 & \text{if } x_i = 0 \\ -1 & \text{if } x_i = 1 \end{cases}$ and with $+$ denoting the element-wise addition of vectors. The discrete difference gradient $\nabla_{SAT}(x)$ is defined for x and all y in $N(x)$ as the vector $\nabla_{SAT}(x, y) = L_d(x, \lambda) - L_d(y, \lambda)$. The discrete difference gradient is the vector of changes of the Lagrangian function for the current λ. The gradient algorithm now follows the direction vector v (moves to the neighbor) which minimizes $L_d(y, \lambda)$ for all $y \in N(x)$, ties are broken arbitrarily.

The gradient algorithm updates $x \in \{0, 1\}^n$ and $\lambda \in R^m$ until a saddle point (fixed point) is reached (this means that $x(k+1) = x(k)$ and $\lambda(k+1) = \lambda(k)$ hold) by the following update rules $x(k+1) = x(k)+v$ and $\lambda(k+1) = \lambda(k)+c \cdot U(x(k))$ where c is a parameter which controls the speed of the increase of the weights of the Lagrangian multipliers over iterations.

In the following, we adapt two concepts of discrete Lagrangian methods for the SAT problem to modularity optimization: The definition of the discrete manifold and the construction of the discrete difference gradient. Since $Q(G, C) \in [-0.5, 1]$, the discrete manifold for modularity maximization is $\Omega \times [-0.5, 1]$.

Figure 11 shows an intermediate partition of a graph clustering problem: each node represents a cluster labelled with the number of edges within the cluster, the edges are weighted with the number of edges between the clusters. Figure 11 visualizes the join operations as a graph rewriting operation which substitutes clusters i and j with a new node and the edge set which results from the graph substitution and a proper update of the numbers of edges within and between clusters.

Let $C^{p-1} = join(C^p, i, j) = join(C^p, j, i)$ be the commutative operation of joining the two clusters C_i and C_j of C^p. The change in modularity from such a join is $\Delta Q(C^p, i, j) = e_{ij} + e_{ji} - 2a_i a_j = 2(e_{ij} - a_i a_j)$. Note, that $\Delta Q(C^p, i, j)$ depends on the partition C^p at which it is evaluated. The path-dependence leads

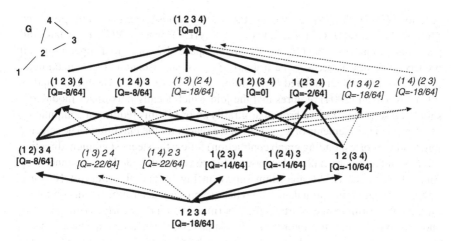

Fig. 12 Path space for graph G (with modularity Q)

to an unbalanced cluster growth (see Ovelgönne and Geyer-Schulz 2012a). From a partition C^p, a join can generate at most $p(p-1)/2$ new partitions with $p-1$ clusters.

Hierarchical agglomerative modularity clustering algorithms start with the n vertices of the graph as singleton clusters ($p = n$) and apply the join operation until a single cluster ($p = 1$) is reached. The sequence of $n-1$ join operations is called a path of the algorithm which consists of n partitions. The path space is the set of all possible paths of the algorithm. The path space structure induced by the join operation defined above adds a lattice structure to Ω. All paths leading to the same partition are equivalent with regard to modularity, because $Q(C^{p-1}) = Q(C^p) + \Delta Q(C^p, i, j)$ holds. For example (for G and the path space shown in Fig. 12), $Q(G, (1\,2\,3)\,4) = Q(G, 1\,2\,3\,4) + \Delta Q(1\,2\,3\,4, 1, 2) + \Delta Q((1\,2)\,3\,4, (1\,2), 3) = Q(G, 1\,2\,3\,4) + \Delta Q(1\,2\,3\,4, 1, 3) + \Delta Q((1\,3)\,3\,4, (1\,3), 2) = Q(G, 1\,2\,3\,4) + \Delta Q(1\,2\,3\,4, 2, 3) + \Delta Q(1\,(2\,3)\,4, (2\,3), 1)$.

Figure 12 shows the path space for a graph G. Because of the observation that the join of two unconnected (by a direct link) vertices (or communities) decreases the modularity, we can discard all joins (dotted arcs in Fig. 12) between unconnected graph components (in italics in Fig. 12). In addition, note that as the examples from Fig. 12 show, paths with a local decrease may still reach the globally optimal solution (Q(−12/64, 1 2 3 4), Q(−14/64, 1 2 (3 4)), Q(0, (1 2) (3 4))) or may be the only way to reach a partition (e.g. by Q(−12/64, 1 2 3 4), Q(−14/64, 1 2 (3 4)), Q(−2/64, 1 (2 3 4))).

The discrete (join) gradient of a partition C^p is $\nabla_{join}(C^p)$. It is the vector of the modularity changes $\Delta Q(C^p, i, j)$ of all possible joins $join(C^p, i, j)$ at C^p. The modularity of the join neighborhood $N_{join}(C^p)$ (a vector with $p(p-1)/2$ components) is given by $Q(N_{join}(C^p)) = Q(G, C^p)\mathbf{1} + \nabla_{join}(C^p)$ with $\mathbf{1}$ a vector of ones of appropriate length. The discrete gradient at a partition C^p is $\nabla(C^p)$ and it is the vector whose elements are given by $Q(G, (C^y)) - Q(G, C^p)$ for all $C^y \in N(C^p)$ where $N(C^p)$ contains all partitions connected to C^p in the path space by one join operation (all predecessor and successor partitions). C^p is an isolated local maximum, if $\nabla(C^p) < 0$, and an isolated local minimum, if $\nabla(C^p) > 0$. The greedy update rule of Newman (2004) requires that the join which locally maximizes modularity should be selected (ties are broken arbitrarily). For C^p this requires a computation of all $p(p-1)/2$ elements of $\nabla_{join}(C^p)$ and a search for the maximum. The selected join (and the corresponding graph contraction) follow the steepest gradient direction (greatest increase or smallest decrease). This corresponds exactly to the inner loop (lines 3–7) of the GSAT algorithm shown in Fig. 9. The algorithm of Newman (2004) is locally join optimal and, if no ties are present, deterministic. In contrast to the GSAT algorithm, no restarts of the algorithm at a random partition are possible, and as a consequence, the algorithm of Newman (2004) can not escape local optima.

The (basic) randomized greedy update rule requires that one cluster of the partition is randomly chosen and that only the joins of this cluster with all other clusters are considered for contraction by a join. For C^p this requires the computation of at most $p - 1$ elements of $\nabla_{join}(C^p)$. The randomized greedy update rule always selects one join partner randomly and then searches the best cluster to join with this partner. This corresponds exactly to the random selection of an unsatisfied clause in line 5 of the WALKSAT algorithm shown in Fig. 10. The analysis of the WALKSAT algorithm has shown that this randomization combined with the Metropolis step is essential for the excellent scalability and performance of the WALKSAT algorithm (Biere et al. 2009, p. 188). Because of its post-processing (RG_Refine), it is a non-deterministic locally 1-change optimal update rule.

We get a family of randomized greedy update rules (see RG_BuildDendrogram algorithm shown in Fig. 4) by adapting the scope of the search by randomly selecting k out of $p-1$ join partners. Both update rules described above belong to this family.

The main advantages of the randomized greedy update rule are: First, a lower computational complexity ($O(n)$ vs. $O(n^2)$ with $n = | V |$), second, a random exploration of local modularity optima, and third, a more balanced cluster growth.

4 From Local to Global Optima: Morse Theory and Core Groups

But how do we move from a local optimum towards a global optimum? The RG algorithm is a non-deterministic local hill-climber. Therefore, the first answer is that by repeated execution of the RG algorithm a sample of local optima can be drawn – all that can be reached from the start partition. Second, for an infinite number of restarts from a randomly selected partition the algorithm reaches the global optimum with a probability of 1. For a discussion of this idea in the context of genetic algorithms see Geyer-Schulz (1992). Another classic is the combination of

Fig. 13 The graph of a
nonlinear function with 2
maxima and 1 saddle
point in R^2
(see Jongen et al. 2000, p. 15)

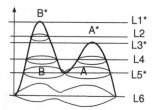

the gradient step with Metropolis moves controlled by an annealing heuristic. With
a very slow annealing heuristic, convergence of the algorithm to a global optimum
is guaranteed (Hajek 1988). All of these approaches lead to algorithms which have
been shown not to scale very well for modularity clustering.

In Ovelgönne and Geyer-Schulz (2010), Michael Ovelgönne introduced the idea
of combining a set of locally optimal partitions to a core group partition as a way
to find restart points for the RG algorithm. This idea has been implemented for
an ensemble learning algorithm in Ovelgönne and Geyer-Schulz (2012b) and it is
responsible for the increase in optimization quality necessary to win the DIMACS
quality challenge. A first interpretation of core group partitions as saddle points
in the partition lattice is presented in Ovelgönne and Geyer-Schulz (2013). In the
following, we give an extended, but still informal explanation of core groups as
high-dimensional saddle points in the partition lattice Ω.

For this purpose, we consider the topology of the global search space with the
help of Morse theory, the study of the behavior of lower level sets of functions as
the level varies (Jongen et al. 2000). M is an open subset of R^n, and $f(M) \to R$ a
function from M to R. By $C^k(M, R)$ we denote the space of k-times continuously
differentiable functions on M and $C^\infty(M, R) = \cap_{k \in N} C^k(M, R)$, where $N =
\{0, 1, 2, \dots\}$. Let f be a function with n variables with $f \in C^\infty(R^n, R)$, f
nondegenerate, and $f(x) = \sum_{i=1}^{n} x_i^2$ for $\| x \| \leq 1$ (Jongen et al. 2000, p. 9).
The lower level sets of f are then defined as $f^\alpha = \{x \in R^n \mid f(x) \leq \alpha\}$ for all
levels $\alpha \in R$ (Jongen et al. 2000, p. 14). Whenever the level passes a stationary or
Karush-Kuhn-Tucker point (that is a local minimum, local maximum, or a saddle
point) the topology of the lower level set changes.

When we move from level L6 to level L4 in Fig. 13, then from level L6 to level
L5 the lower level set is continuously deformed. At level L5 we find the saddle
point, the lower level set breaks in two components linked by the saddle point
(the topological change). And from L5 to L4 the two components are continuously
deformed. A gradient algorithm starting from a point in A runs to A^*, and from
a point in B to B^*. The topologically separate components indicate the basins
of attraction A and B for the gradient algorithm. A non-deterministic gradient
algorithm starting from the saddle point will climb either to A^* or to B^*. Saddle
points are thus the points where the paths of gradient algorithms to local optima
split.

In 2 dimensions, the analysis of the topological structure of the nonlinear space
becomes more complex, as the level set for a bounded nonlinear function with

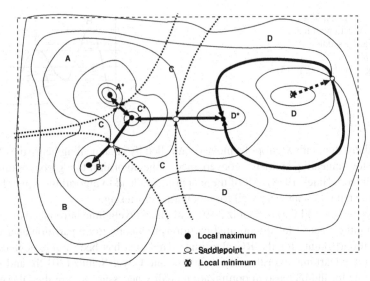

Fig. 14 Level sets of a nonlinear function f in R^2 (see Jongen et al. 2000, p. 10). The *broken arrows* indicate the trajectories from local minima to the saddle points (the *broken arrow* from the local minimum in D to a saddle point). The *full arrows* the trajectories to the local maxima A^*, B^*, C^*, and D^* (rough sketch). We call this graph a Morse graph. The *dotted arrows* separate the (open) basins of attraction A, B, C, and D (and are not part of the Morse graph)

four local maxima, four saddle points, and one local minimum shows (see Fig. 14). However, when we look at the trajectories leading from a saddle point to a local maximum in Fig. 14, at saddle points, the basins of attraction for gradient algorithms are glued together. At saddle points, the paths to local maxima split. A Morse graph is the graph that connects all critical points of a nonlinear function. For example, in Fig. 14 the Morse graph (broken and full arrows) connects all critical points of the nonlinear function shown.

If the number of local maxima of a nonlinear function within a given region (indicated in Fig. 14 as dotted box) is known and finite (say k), we can in principle find the global optimum by sampling: We start a local gradient algorithm from a randomly chosen point, until we have found k local maxima. The largest local maximum is the global maximum (this is not necessarily unique). Sampling works best, if the areas of the basins of attraction of each local maximum are of equal size. However, even in this setting, sampling will be problematic, if the area of the basin of attraction of a local maximum is 0 or near 0. However, in general (without knowing the number of local maxima), this approach will not work as a counter example by Hassler Whitney shows.

Figure 15 shows three situations in which a local minimum or a saddlepoint is a promising start point for a randomized gradient method: The local minimum in the "smooth volcano" situation (Fig. 15a) is a start point from which infinitely many local maxima can be reached. In the "rugged volcano" situation (Fig. 15b), a possibly large, but finite set of local maxima can be reached. The "rugged

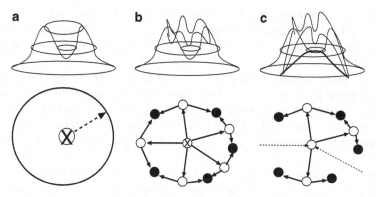

Fig. 15 A smooth volcano (**a**), a rugged volcano (**b**), and a rugged mountain saddle (**c**) together with their Morse graphs

mountain saddle" (Fig. 15c) characterizes a saddle point as a promising start point for the RG algorithm shown in Fig. 1. In all Morse graphs shown in Fig. 15, saddle points split the paths to local maxima. The last two situations in which many local maxima are reachable from a single starting point are characteristic of modularity maximization for most graphs too, because this is a consequence of the large number of graph automorphisms present in many real-life networks (e.g. VLSI circuits, the roadnetworks of Florida and California, the North-American Internet backbone graph) (Sakallah 2009, p. 309). As an illustration of a graph automorphism, consider graph G shown in Fig. 12: The two partitions 1 (2 3) 4 and 1 (2 4) 3 have the same modularity, because the clusters (2 3) and (2 4) are the two isomorphic 2-element subgraphs of the cyclic subgraph G_1 of G with the vertices 2, 3, and 4. We leave it to the reader to discover the other graph automorphisms present in G.

Finally, we observe that when we have a sample of 2 or **more** local maxima, promising starting points are the local minima or saddle points of the minimal spanning tree which connects the sample points.

We can embed the path space of a modularity maximization problem defined in the previous section and shown in Fig. 12 into an appropriate C^∞ space in such a way that the critical points in path space (local maxima, local minima, and the saddle points) are preserved in the continuous space. This embedding establishes a homotopy equivalence between the path-space of modularity maximization and a C^∞ space which preserves the connectedness structure of the critical points on the Morse graph.

In this setting, core group partitions are saddle points:

A core group partition is by its very construction (see Eq. 5) a partition from which all local optima from which it has been generated are reachable by possibly repeatedly applying the join operation of an agglomerative hierarchical clustering algorithm. Because of the join lattice structure of the path space (e.g. for G shown in Fig. 12 and the join operation of the RG algorithm) which is a sublattice of the complete path space (e.g. for G shown in Fig. 3), a core group partition always exists. The level of the core group partition (defined as the number of joins needed to construct it) will always be strictly lower than the lowest level of the maxima it is generated from.

A core group partition corresponds to a saddle point: In the path space of G the core group partition is a branching point, where the join-paths of a gradient algorithm leading to the k local maxima it has been generated from diverge. The core group partition clearly can be reached from the singleton partition (the infimum) of the path-space lattice by a sequence of join operations of a gradient algorithm.

Because a core group partition is a saddle point, it is a good restart point for a RG algorithm, if it contains join-paths to more local optima than it has been generated from. Empirically, setting the number of local maxima to $k = ln(n)$ where n is the number of vertices has worked quite well (see Ovelgönne and Geyer-Schulz 2012b).

To summarize:

- The operation of forming core group partitions from sets of locally maximal (or almost maximal) partitions identifies (some) saddle points (critical points) on the lattice of partitions.
- Core group partitions help in exploring the Morse graph of critical points.
- Core group partitions are good points for restarting RG algorithms, because a core group partition is a branching point in the search space where different basins of attraction meet.

5 Summary and Further Research

In this paper we have analyzed the RG algorithm with the CGGC-scheme and we describe an analogy between the discrete problem of modularity optimization and nonlinear optimization in finite dimensions. We have shown that core group partitions are the discrete counter-parts of saddle-points and that they constitute good restart points for the RG-algorithm.

The behavior of the RG-algorithm mimics the idea to extract the minimal spanning tree of the Morse graph in continuous space. However, what remains to be done, is to construct an algorithm which allows a systematic exploration of the Morse graph of the modularity maximization problem and thus guarantees finding the global maximum of the modularity optimization problem. In addition, in this paper we have not exploited any properties of the automorphism group of the underlying graph.

References

Adamic LA, Glance N (2005) The political blogosphere and the 2004 U.S. election: divided they blog. In: Proceedings of the 3rd international workshop on link discovery, LinkKDD'05, Chicago. ACM, New York, pp 36–43

Agarwal G, Kempe D (2008) Modularity-maximizing graph communities via mathematical programming. Eur J Phys B 66:409–418

Aloise D, Caporossi G, Hansen P, Liberti L, Ruiz M (2012) Modularity maximization in networks by variable neighborhood search. In: 10th DIMACS implementation challenge – graph partitioning and graph clustering. http://www.cc.gatech.edu/dimacs10/papers/[11]-VNS_DIMACS.pdf

Biere A, Heule M, van Maaren H, Walsh T (eds) (2009) Handbook of satisfiability. Frontiers in artificial intelligence and applications, vol 185. IOS, Amsterdam

Boguñá M, Pastor-Satorras R, Díaz-Guilera A, Arenas A (2004) Models of social networks based on social distance attachment. Phys Rev E 70(5):056,122

Boldi P, Rosa M, Santini M, Vigna S (2011) Layered label propagation: a multiresolution coordinate-free ordering for compressing social networks. In: Proceedings of the 20th international conference on world wide web, Hyderabad. ACM, New York

Brandes U, Delling D, Gaertler M, Görke R, Hoefer M, Nikoloski Z, Wagner D (2007) On finding graph clusterings with maximum modularity. In: Graph-theoretic concepts in computer science. Springer, Berlin/New York, pp 121–132

Brandes U, Delling D, Gaertler M, Görke R, Hoefer M, Nikoloski Z, Wagner D (2008) On modularity clustering. IEEE Trans Knowl Data Eng 20(2):172–188

Duch J, Arenas A (2005) Community detection in complex networks using extremal optimization. Phys Rev E 72(2):027,104

Geisberger R, Sanders P, Schultes D (2008) Better approximation of betweenness centrality. In: 10th workshop on algorithm engineering and experimentation, San Francisco. SIAM, pp 90–108

Geyer-Schulz A (1992) On learning in a fuzzy rule-based expert system. Kybernetika 28:33–36

Goldberg A (1979) On the complexity of the satisfiability problem. In: 4th workshop of automatic deduction, Austin, pp 1–6

Guimerà R, Danon L, Díaz-Guilera A, Giralt F, Arenas A (2003) Self-similar community structure in a network of human interactions. Phys Rev E 68(6):065,103

Hajek B (1988) Cooling schedules for optimal annealing. Math Oper Res 13(2):311–329

Jongen H, Jonker P, Twilt F (2000) Nonlinear optimization in finite dimensions. Kluwer Academic, Dordrecht

Kautz H, Sabharwal A, Selman B (2009) Incomplete algorithms. In: Biere A, Heule M, van Maaren H, Walsh T (eds) Handbook of satisfiability. Frontiers in artificial intelligence and applications, vol 185. IOS, Amsterdam, chap 6, pp 185–203

Newman MEJ (2001) The structure of scientific collaboration networks. Proc Natl Acad Sci USA 98(2):404–409

Newman MEJ (2004) Fast algorithm for detecting community structure in networks. Phys Rev E 69(6):066,133

Newman MEJ (2006) Finding community structure in networks using the eigenvectors of matrices. Phys Rev E 74(3):036,104

Newman MEJ, Girvan M (2004) Finding and evaluating community structure in networks. Phys Rev E 69(2):026,113

Ovelgönne M, Geyer-Schulz A (2010) Cluster cores and modularity maximization. In: Fan W, Hsu W, Webb GI, Liu B, Zhang C, Gunopulos D, Wu X (eds) 10th IEEE international conference on data mining workshops, ICDMW'10, Sydney. IEEE Computer Society, Los Alamitos, pp 1204–1213

Ovelgönne M, Geyer-Schulz A (2012a) A comparison of agglomerative hierarchical algorithms for modularity clustering. In: Gaul W, Geyer-Schulz A, Schmidt-Thieme L, Kunze J (eds)

Proceedings of the 34th conference of the German Classification Society, Karlsruhe. Studies in classification, data analysis, and knowledge organization. Springer, Heidelberg, pp 225–232

Ovelgönne M, Geyer-Schulz A (2012b) An ensemble-learning strategy for graph-clustering. In: Bader DA, Meyerhenke H, Sanders P, Wagner D (eds) 10th DIMACS implementation challenge – graph partitioning and graph clustering, Rutgers University. http://www.cc.gatech. edu/dimacs10/papers/[18]-dimacs10_ovelgoennegeyerschulz.pdf

Ovelgönne M, Geyer-Schulz A (2013) An ensemble learning strategy for graph clustering. In: Bader DA, Meyerhenke H, Sanders P, Wagner D (eds) Graph partitioning and graph clustering. Contemporary mathematics, vol 588. American Mathematical Society, Providence, pp 187–205

Ovelgönne M, Geyer-Schulz A, Stein M (2010) Randomized greedy modularity optimization for group detection in huge social networks. In: 4th ACM SNA-KDD workshop on social network mining and analysis, Washington, DC

Sakallah KA (2009) Symmetry and satisfiability. In: Biere A, Heule M, van Maaren H, Walsh T (eds) Handbook of satisfiability. Frontiers in artificial intelligence and applications, vol 185. IOS, Amsterdam, chap 10, pp 289–338

Schuetz P, Caflisch A (2008) Efficient modularity optimization by multistep greedy algorithm and vertex mover refinement. Phys Rev E 77:046,112

Shang Y, Wah BW (1998) A discrete Lagrangian-based global-search method for solving satisfiability problems. J Glob Optim 12(1):61–99

Smale S (1983) On the average number of steps of the simplex method of linear programming. Math Program 27(3):241–262

Watts DJ, Strogatz SH (1998) Collective dynamics of "small-world" networks. Nature 393(6684):440–442

Whitney H (1935) A function not constant on a connected set of critical points. Duke Math J 1:514–517

Zhu Z, Wang C, Ma L, Pan Y, Ding Z (2008) Scalable community discovery of large networks. In: Proceedings of the 2008 ninth international conference on web-age information management, WAIM'08, Zhangjiajie, IEEE Computer Society, Los Alamitos, pp 381–388

Comparison of Two Distribution Valued Dissimilarities and Its Application for Symbolic Clustering

Yusuke Matsui, Yuriko Komiya, Hiroyuki Minami, and Masahiro Mizuta

Abstract There are increasing requirements for analysing very large and complex datasets derived from recent super-high cost performance computer devices and its application software. We need to aggregate and then analyze those datasets. Symbolic Data Analysis (SDA) was proposed by E. Diday in 1980s (Billard L, Diday E (2007) Symboic data analysis. Wiley, Chichester), mainly targeted for large scale complex datasets. There are many researches of SDA with interval-valued data and histogram-valued data. On the other hand, recently, distribution-valued data is becoming more important, (e.g. Diday E, Vrac M (2005) Mixture decomposition of distributions by copulas in the symbolic data analysis framework, vol 147. Elsevier Science Publishers B. V., Amsterdam, pp 27–41; Mizuta M, Minami H (2012) Analysis of distribution valued dissimilarity data. In: Gaul WA, Geyer-Schulz A, Schmidt-Thieme L, Kunze J (eds) Challenges at the interface of data analysis, computer science, and optimization. Studies in classification, data analysis, and knowledge organization. Springer, Berlin/Heidelberg, pp 23–28). In this paper, we focus on distribution-valued dissimilarity data and hierarchical cluster analysis. Cluster analysis plays a key role in data mining, knowledge discovery, and also in SDA. Conventional inputs of cluster analysis are real-valued data, but in some cases, e.g., in cases of data aggregation, the inputs may be stochastic over ranges, i.e., distribution-valued dissimilarities. For hierarchical cluster analysis, an order relation of dissimilarity is necessary, i.e., dissimilarities need to satisfy the properties of an ultrametric. However, distribution-valued dissimilarity does not have a natural order relation. Therefore we develop a method for investigating order relation of

Y. Matsui (✉)
Graduate School of Information Science and Technology, Hokkaido University, Sapporo, Japan
e-mail: matsui@iic.hokudai.ac.jp

Y. Komiya · H. Minami · M. Mizuta
Information Initiative Center, Hokkaido University, Sapporo, Japan
e-mail: komiya@iic.hokudai.ac.jp; min@iic.hokudai.ac.jp; mizuta@iic.hokudai.ac.jp

W. Gaul et al. (eds.), *German-Japanese Interchange of Data Analysis Results*,
Studies in Classification, Data Analysis, and Knowledge Organization,
DOI 10.1007/978-3-319-01264-3__3,
© Springer International Publishing Switzerland 2014

distribution-valued dissimilarity. We also apply the ordering relation to hierarchical symbolic clustering. Finally, we demonstrate the use of our order relation for finding a hierarchical cluster of Japanese Internet sites according to Internet traffic data.

1   Introduction

Recent abundant computer resources make it easy to collect data of large scale and complex structure like *Bigdata*. This dramatic fall of the cost performance ratio of computing brings us more possibilities of discovering new knowledge, but at the same time we are facing a lot of problems of how to analyze such large scale complex datasets. We need methods reducing scale and complexity but keeping information quantity and quality in balance and using as much of the information as possible in our analysis.

1.1   Symbolic Data Analysis

Symbolic Data Analysis (SDA) has been proposed by E. Diday in 1980s (Billard and Diday 2007). In conventional multivariate data analysis, a $n \times p$ data matrix is represented as n points in p dimensions. On the other hand, in SDA, we often use interval-valued data. In this case, the data can be represented as m ($< n$) hyperrectangles (called *symbolic objects*) in p dimensions. Points are uniformly distributed within each hyperrectangles. Then symbolic object has internal variations.

A symbolic object is also said to be a second-level object, i.e., category, class, or concept. For instance, *the* dog indicates just one observation (i.e., first-level object), but if we are interested in *a* dog in a sense of one species, it is a second-level object and *a* dog may include a huge number of dogs. In SDA, we use descriptions for objects.

We can consider many possible data descriptions for symbolic objects, e.g., interval values, modal interval values, and distribution values. Interval value is as $[10, 30]$, modal-interval value is as $\{[10, 20); 0.3, [20, 30); 0.7\}$. At the same time, however, there is a need to develop methods for expanding the treatment of conventional real values into symbolic treatments, and it is also a need for extension of the multivariate analysis to be suitable with in SDA. For instances, principal component analysis, multidimensional scaling(MDS) and cluster analysis etc. has been developed for interval-valued data.

1.2   Distribution-Valued Data

In this paper, we are interested in distribution-valued data. The distribution-valued representation is a powerful description for large, complex datasets. Interval values can be an efficient description for large complex data, but this description assumes to take values *uniformly* over the interval range, so that there is a loss w.r.t. information from the original data. Distribution-valued data is a generalization of interval-valued data in the sense that distribution-valued data can represent arbitrary distributions of values. The recent advances in SDA are concerned with distribution-valued data (e.g. (Diday and Vrac 2005), (Mizuta and Minami 2012) etc.).

1.3 Dissimilarity Data Represented by Distribution-Values

We focus on dissimilarities given by the distribution-values. Dissimilarity data often rises as input data in the context of MDS and cluster analysis. Generally, when we use measured dissimilarity data, we presume error, so that we take average on the data to use dissimilarity. However, since the dispersion is not always error, it is also natural to consider to use the all values for analysis. We regard observations from trials as distribution-valued data, and analyze them directly.

2 Hierarchical Cluster Analysis for Distribution-Valued Dissimilarities Data

In this section, we propose the new method of hierarchical cluster analysis for distribution-valued dissimilarity data. The targeted data of SDA is *concepts* instead of *individuals* in conventional multivariate analysis. We assume that there are n concepts, and input data is distribution-valued dissimilarity. At first we give a quick review of conventional hierarchical cluster analysis.

Hierarchical cluster analysis is represented mathematically as n-tree as nested tree structure model. Let C be an arbitrary set to be clustered and $\{\phi\}$ be empty set. Then n – tree is a set τ of subsets on C satisfying: $C \in \tau; \phi \notin \tau; \{i\} \in \tau$ for every $i \in C$; and $A \cap B \in \{\phi, A, B\}$ for all $A, B \in \tau$ (McMorris and Neumann 1983; Gordon 1985, 1987) etc.

A dendrogram is an n-tree with the additional property that height h is associated with each subset, satisfying the condition $A \cap B \neq \phi, h(A) \leq h(B) \Leftrightarrow A \subseteq B$ for subsets $A, B \in \tau$. A necessary and sufficient condition for a dendrogram is the following;

$$h(\{i, j\}) \leq min\{h(\{j, k\}), h(\{i, k\})\} \quad \forall i, j, k \in C.$$

This condition is called the *ultra metric* condition. An ultrametric condition requires, in other words, that sets of dissimilarities must have (weak) order relations. This point is one of the main themes in this paper. In general, distribution-valued dissimilarities don't have a natural order like real values. The next following sections, we give the a new method for ordering distribution values.

2.1 Comparison of Two Distributions and Ordering Distributions

We start with considering how to compare two distribution values, i.e., which distribution value is larger, otherwise they are equivalent. Maybe the first idea for this question is using mean values of each distribution – compare the mean values and reflect their order relation. This method (say, *mean method* in this paper) is simple, but there is much loss of information from the original data. In the following sections, we develop the method to compare two distributions which includes more information and respects, e.g., variations and asymmetry. After that we extend the method to establish an order relations among dissimilarities.

2.1.1 Related Work

Stochastic dominance is one of notable works on comparing distributions, where first-order dominance and second-order dominance are defined (Levy 2006). First-order dominance is defined by the rule $X \succ Y \Leftrightarrow F(t) \leq G(t)$ for all t, where X, Y are random variable and $F(t) = \Pr(X \leq t)$, $G(t) = \Pr(Y \leq t)$. Second-order dominance is defined by the rule of $X \succ Y \Leftrightarrow \int_{-\infty}^{u} F(t)dt \leq \int_{-\infty}^{u} G(t)dt$ for all u. Necessary condition of first-order dominance and second-order dominance is $E(X) \geq E(Y)$.

Thus stochastic dominance is comparing random variables through comparison of distribution functions and its integral values pointwise. Our distribution-valued data approach is to directly compare random variables, i.e., using the value of $\Pr(X \leq Y)$, based on comparison of observations. Besides, our approach does not necessarily need the condition $E(X) \geq E(Y)$ even when X is larger than Y. We give basic definitions in the next section.

2.1.2 Definitions and New Methods

Definition by Probability

We put S_{ij} as dissimilarity with distribution values between concept i and j, where $S_{ij} = S_{ji}$. Note that S_{ij} is a distribution value. s_{ij} is a random variable of distribution S_{ij}. Now we focus on the concepts i, j, k, and l, i.e. dissimilarities S_{ij} and S_{kl}. We define

$$\Pr(s_{ij} > s_{kl}) \equiv P_{ij,kl}. \tag{1}$$

If S_{ij} is analytical, we can calculate $P_{ij,kl}$ as follows.

$$\Pr(s_{ij} > s_{kl}) = \Pr(s_{ij} - s_{kl} > 0) = \Pr(s'_{ij,kl} > 0) = \int_{0}^{s'_{ij}} S_{ij} * S_{kl} ds' \tag{2}$$

where $s_{ij} - s_{kl} = s'_{ij,kl}$ and '$*$' means convolution. Here we define the rule for comparison of two distributions.

Definition 1 (Comparison of two distributions). A comparison operator $\overset{(\infty)}{\succ}$ for $\{S_{ij}\}$ is defined as

$$S_{ij} \overset{(\infty)}{\succ} S_{kl} \text{ iff } P_{ij,kl} > \frac{1}{2}. \tag{3}$$

In general, a transitive relation among the distributions does not hold. Then we introduce the value taking summation over k, l, i.e., $P_{ij,\cdot\cdot} = \frac{1}{N} \sum_{k,l=1}^{N} P_{ij,kl}$. We can interpret $P_{ij,\cdot\cdot}$ as probability that S_{ij} is larger than any other dissimilarity. Based on the comparison operator $\overset{(\infty)}{\succ}$, i.e., $P_{ij,\cdot\cdot}$, we define comparison operator $\overset{(\infty)^\sigma}{\succ}$.

Definition 2 (Order relation among distributions). Another comparison operator $\overset{(\infty)^\sigma}{\succ}$ for $\{S_{ij}\}$ is defined as

$$S_{ij} \overset{(\infty)^\sigma}{\succ} S_{kl} \text{ iff } P_{ij,\cdot\cdot} > \frac{1}{2}. \tag{4}$$

Definition by Sample Sets

The previous definitions work, if the probability distributions S_{ij} is known. However, in practical situations, we cannot get distribution functions but sample only. Then, we propose definitions for sample: Definitions 3 and 4. At first, we begin with the approximated method for comparison of two distributions. The basic algorithm of the comparison method is as follows.

1. Set initializing; set the number of trials M; set $m \leftarrow 0$.
2. Trial: Sample randomly with replacement from each distribution.
3. Compare the two values from the trial.
4. Give one point to the distribution with larger value, 0.5 point to that with equal value, otherwise zero. $m \leftarrow m + 1$.
5. Repeat from 2 to 4 until $m = M$.
6. Count all points of each distribution, the larger one is output.

Basic idea of this algorithm is like this: If we observe every trial that which of the sample value is larger or smaller, then counting these results (points), we can empirically guess which of distribution-value is larger or smaller.

We formulate the algorithm above. We put observations of random variables s_{ij}, s_{kl} as s_{ij}^*, s_{kl}^*. Then we define the sequence of numbers $X_{ij,kl}^l; l = 1, 2, \cdots, M$ as follows.

$$X^l_{ij,kl} = \begin{cases} 1 & (if \quad s^*_{ij} > s^*_{kl}) \\ \frac{1}{2} & (if \quad s^*_{ij} = s^*_{kl}) \\ 0 & (otherwise) \end{cases}. \tag{5}$$

$P_{ij,kl}$ is empirically estimated as

$$P^*_{ij,kl} = \frac{1}{M} \sum_{l=1}^{M} X^l_{ij,kl}.$$

We replace $P_{ij,kl}$ on $P^*_{ij,kl}$ in Definition 1, then we can define a rule for an empirical version of comparison of two distributions.

Definition 3 (Comparison of two distributions based on sample sets). A comparison operator $\overset{\text{(freq)}}{\succ}$ for $\{S_{ij}\}$ is defined as

$$S_{ij} \overset{\text{(freq)}}{\succ} S_{kl} \quad \text{iff} \quad P^*_{ij,kl} > \frac{1}{2}. \tag{6}$$

As in Definition 1, $\overset{\text{(freq)}}{\succ}$ is the order relation of the two, however, in this time, based on empirical frequency. In the same way, We define order relations among the distributions.

For transitive relation among distributions, we put $P^*_{ij,..} = \frac{1}{N} \sum_{k,l=1}^{N} P^*_{ij,kl}$. We can interpret $P^*_{ij,..}$ as probability that S_{ij} is larger than any other dissimilarity. Based on the comparison operator $\overset{\text{(freq)}}{\succ}$, i.e., $P^*_{ij,..}$, we define comparison operator $\overset{\text{(freq)}^{\sigma}}{\succ}$.

Definition 4 (Order relation among distributions based on sample sets). Another comparison operator $\overset{\text{(freq)}^{\sigma}}{\succ}$ for $\{S_{ij}\}$ is defined as

$$S_{ij} \overset{\text{(freq)}^{\sigma}}{\succ} S_{kl} \quad \text{iff} \quad P^*_{ij,..} > \frac{1}{2}. \tag{7}$$

2.2   Extension to Hierarchical Cluster Analysis

We apply the method of comparing and ordering the distributions to hierarchical cluster analysis. There are mainly two parts in hierarchical cluster analysis, i.e., choose the pair of clusters to be merged, and update the dissimilarity between the clusters. In the former part, merge two clusters with minimal distribution-valued dissimilarity which is chosen by $\overset{\text{(freq)}^{\sigma}}{\succ}$ operator. On the other hand in the latter part,

we can use the method of comparing the two dissimilarities then we use $\overset{(freq)}{\succ}$ operator.
Here is the algorithm of the hierarchical cluster analysis. Basic idea is the same as
for conventional hierarchical cluster analysis.
Algorithm with the proposed method

1. At an initial state, there are n clusters; $\{\{1\}, \{2\}, \ldots, \{n\}\}$, i.e., regard each
   concepts as clusters.
2. Calculate $P_{ij,kl}$ for all i, j, k, l.
3. Calculate $P_{ij,..}$ for all i, j.
4. Merge the pair (i, j) such that $(i, j) = \text{argmin} \{P_{ij,..}\}$.
5. Update dissimilarity: New cluster is put $h = i \cup j$, then set P_{hl} such that $P_{hl} = \min \{P_{il}, P_{jl}\}$.
6. Repeat from 2 to 5 until the number of clusters is 1.

In step 2, we use the definition 5 to decide the dissimilarity with minimum
order. And step 3, we apply the definition 3 to update the dissimilarities between
the clusters.

3 Application

We offer the application to network traffic data. We often feel that the browsing
speed on the Web is too slow, but never realize the reason. We can investigate the
reasons related to our personal computer, but it must be hard if the true reason is
far from us, i.e., due to the Web server and/or network traffic jam caused by heavy
users in the intermediate nodes from the source to us.

An engineer group had tried to investigate practical efficiency on the Internet
in Japan. From the viewpoint of network engineering, they had to clarify the
configuration from the bottom (physical layer). However, due to business reason,
all Internet Service Providers in Japan hesitated to reveal the information about
their operating networks. Then, the engineer group had tried to collect the traffic
information by their own applications. On a second thought, the end user can touch
the information mostly from the top layer and feel the speed. The group believed
that the collected data should give us a good approximation on the practical network
traffic, not the theoretical one.

Taking the great size and the periodic trend into consideration, the group had
collected the practical traffic data on the top layer with 'ping' command. The
command gives us Round Trip Time (RTT) which means the time from when it
carries out to the issuer gets a reply from the destination node. Eventually, the
collected data are as follows:

Number of Nodes   35
Number of Issued commands per 1 collection   50
Collection Interval   Every 5 min
Collection Period   Over 2 years

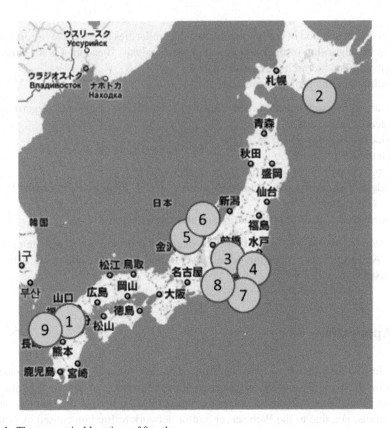

Fig. 1 The geometrical locations of 9 nodes

The group had no way to analyze the big data and need our help. We roughly investigate the data and they are suitable to apply our method, to clarify the relation between the measured nodes.

Unfortunately, the data has tons of missing values and we and the group cannot specify why the operations failed. Then, we use the complete data (without missing values) and the number of target nodes is 9. The left map in Fig. 1 displays the geographical configuration on the 9 nodes.

In the paper, we use the data from 2006.1.18 to 2006.1.21 and each observed datum between the 2 nodes is regarded as distribution valued data in SDA.

Figure 2 shows the clustering results with the conventional way (left) and our proposed one (right). Assuming four clusters from the dendrograms, the classified nodes are displayed in Fig. 3.

During the period, the bad news related to a big IT company is open to the public and it must affects economy in Japan. Node 7 and 8 are known as famous economical news sites and we can assume that Node 7 is isolated in both due to the heaviest traffic. We would persuade ourselves if the node and Node 8 consist of one

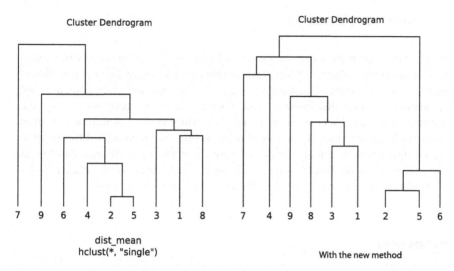

Fig. 2 Results of hierarchical cluster analysis. The *left* dendrogram shows a result of mean method with single-linkage. The *right* one shows a result of the proposed method

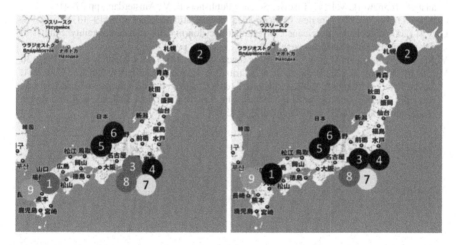

Fig. 3 Partition of nodes in four clusters. The *left* is a result of the mean method, and the *right* is a result of the proposed method

cluster, but the conventional result offers Node 1 is a member of the cluster, too. It is not related to commercial events and might be misclassified from the economical viewpoint.

On the other hand, the proposed result offers that Node 7–9 consist of isolated clusters. Node 9 is in a national university network and needs to pass much routers to Node 7 and 8 (since the academic network has several inter-exchange points for commercial ones in Japan). The others are neither related to commerce nor in academic network. In short, the proposed method gives us a better interpretation rather than the previous ones.

4 Concluding Remarks

In this paper, we handle distribution-valued data. We focus on the dissimilarities with distribution values and hierarchical cluster analysis. In hierarchical cluster analysis, an ultrametric condition is essential. Then we introduce a method for comparison of two distributions, and extend it to order relations among the distribution values. This method is intended to reduce the loss of information from original data, and in fact, we can show that this proposed method is different from the mean method by actual example of network traffic data. We also develop an application of the method to symbolic clustering, and a result of the clustering for traffic data is more interpretable than the case of using mean method.

References

Billard L, Diday E (2007) Symbolic data analysis. Wiley, Chichester

Diday E, Vrac M (2005) Mixture decomposition of distributions by copulas in the symbolic data analysis framework, vol 147. Elsevier Science Publishers B. V., Amsterdam, pp 27–41

Gordon AD (1985, 1987) A review of hierarchical classification. JSTOR 150:119–137

Levy H (2006) Stochastic dominance: investment decision making under uncertainty. Studies in risk and uncertainty. Springer, New York

McMorris FR, Neumann D (1983) Consensus functions defined on trees. Math Soc Sci 4:131–136

Mizuta M, Minami H (2012) Analysis of distribution valued dissimilarity data. In: Gaul WA, Geyer-Schulz A, Schmidt-Thieme L, Kunze J (eds) Challenges at the interface of data analysis, computer science, and optimization. Studies in classification, data analysis, and knowledge organization. Springer, Berlin/Heidelberg, pp 23–28

Pairwise Data Clustering Accompanied by Validation and Visualisation

Hans-Joachim Mucha

Abstract Pairwise proximities are often the starting point for finding clusters by applying cluster analysis techniques. We refer to this approach as pairwise data clustering (Mucha HJ (2009) ClusCorr98 for Excel 2007: clustering, multivariate visualization, and validation. In: Mucha HJ, Ritter G (eds) Classification and clustering: models, software and applications. Report 26, WIAS, Berlin, pp 14–40). A well known example is Gaussian model-based cluster analysis of observations in its simplest settings: the sum of squares and logarithmic sum of squares method. These simple methods can become more general by weighting the observations. By doing so, for instance, clustering the rows and columns of a contingency table will be performed based on pairwise chi-square distances. Finding the appropriate number of clusters is the ultimate aim of the proposed built-in validation techniques. They verify the results of the two most important families of methods, hierarchical and partitional clustering. Pairwise clustering should be accompanied by multivariate graphics such as heatmaps and plot-dendrograms.

1 Introduction

Cluster analysis aims at finding interesting partitions or hierarchies of a set of objects without taking into account any background knowledge. Here, without loss of generality, we consider the cluster analysis of a set of observations (row points of a data table). Model-based cluster analysis has become very popular by the paper of Banfield and Raftery (1993). In the present paper, we focus on Gaussian model-based cluster analysis of observations in its simplest settings that result in the sum of squares and logarithmic sum of squares method. These simple model-based

H.-J. Mucha (✉)
Weierstrass Institute for Applied Analysis and Stochastics (WIAS), D-10117 Berlin, Germany
e-mail: mucha@wias-berlin.de

W. Gaul et al. (eds.), *German-Japanese Interchange of Data Analysis Results*,
Studies in Classification, Data Analysis, and Knowledge Organization,
DOI 10.1007/978-3-319-01264-3_4,
© Springer International Publishing Switzerland 2014

Gaussian clustering techniques can be expressed in terms of clustering based on pairwise squared Euclidean distances between the observations. Moreover, they can be generalized easily by taking into account weights of observations. With respect to the special weights of observations (often referred to as masses), namely the average row profile or row centroid, the cluster analysis of contingency tables based on the chi-square distances results in an optimum decomposition of the Pearson chi-square statistic (Greenacre 1988). The chi-square distance itself is a squared weighted Euclidean distance with the special weights called inverse average column profile. Both hierarchical and partitional cluster analysis of the rows and columns of a contingency table can be carried out based on pairwise χ^2-distances.

In practice, the principle of weighting of observations is a key idea in data mining to deal with both aggregated data (such as cores) and outliers. In the case of outliers one has to down-weight them in order to reduce their influence. Mucha (2004) used special (random) weights for resampling in validation of hierarchical clustering. Or, clustering based on graph theory was generalized by consideration of weights of observations (Bartel et al. 2003).

Hierarchical clustering is in some sense more general than partitional clustering because one gets out a battery of (nested) partitions. Figure 1 shows both the starting point of hierarchical clustering, a distance matrix as a heatmap, and the final hierarchy that is visualised by a dendrogram. On the other hand, partitional clustering results usually in better solutions (criterion values) with respect to a partition into a fixed number of clusters. (A partitional clustering is simply a division of the set of observations into non-overlapping subsets (clusters) such that each data object is in exactly one subset.) Two well-known cluster analysis techniques, *Ward*'s hierarchical method and the partitional K-means clustering, optimize the sum of squares criterion that is derived from the simplest model of K Gaussian sub-populations where we are looking for clusters of the same volume (see for details: Banfield and Raftery 1993). In the present paper, we will consider a "generalization" that can find clusters of different volumes.

Another focus is on validation and visualisation. Our built-in validation techniques (Mucha 2009) can verify the results of the two most important families of methods, the hierarchical and partitional cluster analysis. The Excel-spreadsheet is both a distinguished repository for data/distances/clusters/hierarchies and a plotting board for multivariate graphics that can be provided in Visual Basic for Applications (VBA). Examples are dendrograms, plot-dendrograms, scatterplot matrices, density plots, principal components analysis plots, and correspondence analysis plots (Mucha et al. 2005).

2 Simple Model-Based Gaussian Clustering

Suppose there are K subpopulations with unknown J-dimensional normal densities. Consider a fixed partition $C = (C_1, \ldots, C_K)$ of a sample of I independent observations $\mathbf{x}_i$ into K clusters. The most general model-based Gaussian clustering is when the covariance matrix $\mathbf{W}_k$ of each cluster (subpopulation) C_k is allowed

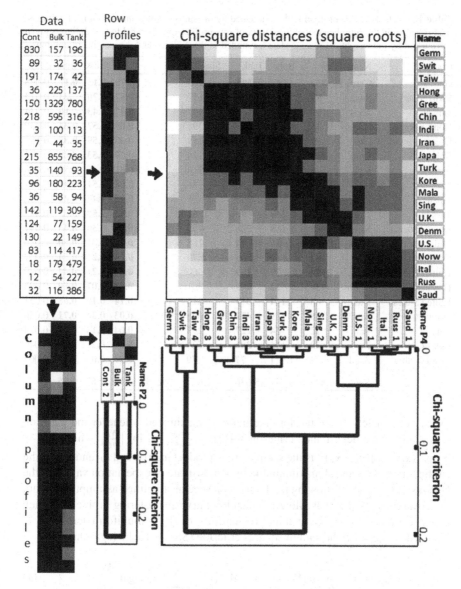

Fig. 1 *Ward*'s hierarchical cluster analysis of Table 1. The *darker* the *gray color* in the heatmap on the *right hand* side the smaller the underlying χ^2-distances are. On the *left hand* side at the *bottom* the cluster analysis of the column points is visualised. (Strictly speaking: *Ward*'s minimum within cluster sum of squares method using both the special weights of rows and the special weights of columns: for details see Sect. 2.2 below)

Table 1 Fleets data. The original table is attached by the sums of rows and the sums of columns. Additionally, on the right hand side, the matrix of row profiles (5) (rounded to two digits) is given surrounded by the weights of rows ("Mass", rounded to three digits) and the weights of columns

Country	Name	Cont	Bulk	Tank	Sum		Cont	Bulk	Tank	Mass
Germany	Germ	**830**	**157**	**196**	1,183		**0.70**	**0.13**	**0.17**	0.098
Switzerland	Swit	**89**	**32**	**36**	157		**0.57**	**0.20**	**0.23**	0.013
Taiwan	Taiw	**191**	**174**	**42**	407		**0.47**	**0.43**	**0.10**	0.034
Hong Kong	Hong	**36**	**225**	**137**	398		**0.09**	**0.57**	**0.34**	0.033
Greece	Gree	**150**	**1,329**	**780**	2,259		**0.07**	**0.59**	**0.35**	0.187
China	Chin	**218**	**595**	**316**	1,129		**0.19**	**0.53**	**0.28**	0.094
India	Indi	**3**	**100**	**113**	216		**0.01**	**0.46**	**0.52**	0.018
Iran	Iran	**7**	**44**	**35**	86		**0.08**	**0.51**	**0.41**	0.007
Japan	Japa	**215**	**855**	**768**	1,838		**0.12**	**0.47**	**0.42**	0.152
Turkey	Turk	**35**	**140**	**93**	268		**0.13**	**0.52**	**0.35**	0.022
Korea (South)	Kore	**96**	**180**	**223**	499		**0.19**	**0.36**	**0.45**	0.041
Malaysia	Mala	**36**	**58**	**94**	188		**0.19**	**0.31**	**0.50**	0.016
Singapore	Sing	**142**	**119**	**309**	570		**0.25**	**0.21**	**0.54**	0.047
United Kingdom	U.K.	**124**	**77**	**159**	360		**0.34**	**0.21**	**0.44**	0.030
Denmark	Denm	**130**	**22**	**149**	301		**0.43**	**0.07**	**0.50**	0.025
United States	U.S.	**83**	**114**	**417**	614		**0.14**	**0.19**	**0.68**	0.051
Norway	Norw	**18**	**179**	**479**	676		**0.03**	**0.26**	**0.71**	0.056
Italy	Ital	**12**	**54**	**227**	293		**0.04**	**0.18**	**0.77**	0.024
Russia	Russ	**32**	**116**	**386**	534		**0.06**	**0.22**	**0.72**	0.044
Saudi Arabia	Saud	**1**	**1**	**77**	79		**0.01**	**0.01**	**0.97**	0.007
	Sum	2,448	4,571	5,036	12,055	Weight	2.22	1.62	1.55	

to vary completely. Then the log-likelihood is maximized whenever the partition C minimizes the determinant criterion $W(C) = \sum_{k=1}^{K} n_k \log |\frac{\mathbf{W}_k}{n_k}|$, where $\mathbf{W}_k = \sum_{i \in C_k} (\mathbf{x}_i - \bar{\mathbf{x}}_k)(\mathbf{x}_i - \bar{\mathbf{x}}_k)^T$ is the sample cross-product matrix, and n_k and $\bar{\mathbf{x}}_k$ are the cardinality and the usual maximum likelihood estimate of expectation values of the kth cluster C_k. In the following the focus is on special simplified assumptions about the covariance matrix that allow to consider pairwise clustering problems. When it is constrained to be diagonal and (a) uniform or (b) non-uniform across all K assumed clusters, (a) the sum of squares criterion (SS) or (b) the logarithmic SS

$$W(C) = \sum_{k=1}^{K} \mathrm{tr}(\mathbf{W}_k) \quad \text{or} \quad W^*(C) = \sum_{k=1}^{K} n_k \log \mathrm{tr}(\frac{\mathbf{W}_k}{n_k}) \tag{1}$$

has to be minimized. Both SS criteria (1) can be reformulated as the minimization of

$$W(C) = \sum_{k=1}^{K} \frac{1}{2n_k} \sum_{i \in C_k} \sum_{h \in C_k} d_{ih} \quad \text{or} \quad W^*(C) = \sum_{k=1}^{K} n_k \log(\sum_{i \in C_k} \sum_{h \in C_k, h>i} \frac{1}{n_k^2} d_{ih}), \tag{2}$$

where $d_{ih} = \|\mathbf{x}_i - \mathbf{x}_h\|^2$ is the pairwise squared Euclidean distance.

2.1 Weighted Observations and Squared Euclidean Distance

Usually, all observations have the same weight. The (logarithmic) SS criteria (1) and (2) can be generalized by weighting the observations. This allows fast and efficient algorithms (analysis of massive data, resampling, . . .) because the distances themselves are not affected by weighting. There are other advantages such as clustering of contingency tables based on the χ^2-distance (see below) or handling of outliers by down-weighting them in some way in order to reduce their influence. Using weights of observations the criteria (2) can be formulated for general use as

$$W(C) = \sum_{k=1}^{K} w(C_k) = \sum_{k=1}^{K} \frac{1}{M_k} \sum_{i \in C_k} m_i \sum_{h \in C_k, h > i} m_h d_{ih} \qquad (3)$$

and

$$W^*(C) = \sum_{k=1}^{K} M_k \log \left(\sum_{i \in C_k} \sum_{h \in C_k, h > i} \frac{m_i m_h}{M_k^2} d_{ih} \right), \qquad (4)$$

where $M_k = \sum_{i \in C_k} m_i$ and $m_i > 0$ denote the weight of cluster C_k and the weight of observation i, respectively. For example, given a partition C, the SS criterion (3) can be improved (minimized) by exchanging an observation i coming from cluster C_k and shifting into cluster C_g if $w(C_k \backslash \{i\}) + w(C_g \cup \{i\}) < w(C_k) + w(C_g)$, where

$$w(C_k \backslash \{i\}) = \frac{1}{M_k - m_i} \left(\sum_{l \in C_k} \sum_{h \in C_k, h > l} m_l m_h d_{lh} - \sum_{h \in C_k} m_i m_h d_{ih} \right)$$

and

$$w(C_g \cup \{i\}) = \frac{1}{M_g + m_i} \left(\sum_{l \in C_g} \sum_{h \in C_g, h > l} m_l m_h d_{lh} + \sum_{h \in C_g} m_i m_h d_{ih} \right).$$

2.2 Hierarchical and Partitional Clustering Based on χ^2-Distance

Clustering techniques can be used to segment a heterogeneous two-way contingency table into smaller, homogeneous parts. Following the paper of Greenacre (1988), here the focus also is on chi-square decompositions of the Person chi-square statistic by clustering the rows and/or the columns of a contingency table. Especially the hierarchical *Ward*'s method as well as its generalization based on (4) is of interest. The latter can find clusters of different volumes. Additionally, one can show that it is also possible to carry out partitional clustering by starting from pairwise chi-square distances. The partitional cluster analysis optimizes directly a criterion with respect to a fixed number of clusters K. Often it attains better solutions than hierarchical cluster analysis.

The ubiquitous starting point of pairwise clustering techniques is a distance matrix $\mathbf{D} = (d_{ih})$. There are at least two well-known families of methods for minimizing the (logarithmic) SS criteria based on pairwise distances: Partitional clustering such as $D_{ih}Ex$ (that is based on the so-called $T_{ih}ExM$ method by Späth 1985) minimizes (3) for a fixed K by exchanging observations between clusters, and hierarchical cluster analysis minimizes (3) in a stepwise manner by agglomeration of pairs of observations/clusters. The well-known K-means method becomes a special case of the $D_{ih}Ex$ method in the framework of pairwise clustering based on squared Euclidean distances without using centroids anymore (see Späth 1985 and its generalization by Bartel et al. (2003)). Concerning hierarchical clustering it should be mentioned that in the case of logarithmic SS usually the process of amalgamation bears decreasing distance levels (inversions). In order to make it possible to draw a dendrogram special treatments are available in our prototype software ClusCorr98 (Mucha 2009).

Now let's consider a contingency table $\mathbf{N} = (n_{ij})$, $i = 1, 2, \ldots, I$, $j = 1, 2, \ldots, J$, as presented in Table 1 in bold on the left hand side. Then

$$\mathbf{x}_i = \mathbf{n}_i / n_{i.} \quad i = 1, 2, \ldots, I \tag{5}$$

are the row profiles (see in Table 1 on the right hand side). Herein $n_{i.} = \sum_{j=1}^{J} n_{ij}$ is the row total of the row point i. That is, we are now back formally to the analysis of a data matrix $\mathbf{X}$. That is, the weighted squared Euclidean distance $d_{ih} = \|\mathbf{x}_i - \mathbf{x}_h\|_Q^2 = \sum_{j=1}^{J} q_j (x_{ij} - x_{hj})^2$ is the appropriate distance measure between the observations i and h, where $\mathbf{x}_i$ and $\mathbf{x}_h$ are the corresponding row profiles and the matrix $\mathbf{Q}$ of weights of variables (column points) is diagonal with the following special weights $q_j = n_{..}/n_{.j}$, where $n_{.j}$ ($j = 1, 2, \ldots, J$) (column total) and $n_{..}$ (grand total) are $n_{.j} = \sum_{i=1}^{I} n_{ij}$ and $n_{..} = \sum_{i=1}^{I} \sum_{j=1}^{J} n_{ij}$, respectively. The dissimilarity between two row profiles $\mathbf{x}_i$ and $\mathbf{x}_h$ can be expressed by using the original contingency table $\mathbf{N}$ as

$$d_{ih} = \sum_{j=1}^{J} \frac{n_{..}}{n_{.j}} \left(\frac{n_{ij}}{n_{i.}} - \frac{n_{hj}}{n_{h.}} \right)^2 , \tag{6}$$

which is the well-known χ^2-distance.

Figure 1 displays both the original data $\mathbf{N}$ and schematically the matrix of row profiles $\mathbf{X}$ that is derived from $\mathbf{N}$. The darker the gray color in the heatmap of the profiles the smaller the underlying values are. The corresponding matrix $\mathbf{D}$ of the pairwise χ^2-distances (6) between the objects (countries) is given also schematically in Fig. 1 on the right hand side at the top. The dendrogram on the right hand side presents the final result of the cluster analysis of the observations based on (3) with the following special weights (masses) $m_i = n_{i.}/n_{..}$ of the observation i, $i = 1, 2, \ldots, I$. The hierarchical clustering of variables is shown below on the left hand side.

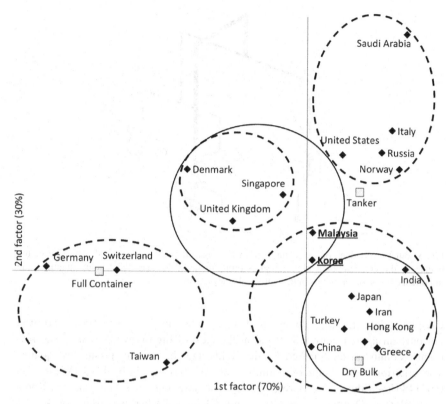

Fig. 2 Correspondence analysis (CA) plot of row and column points of Table 1. In addition, both a partition of *Ward*'s method is marked by ellipses (*dashed lines*) and the cluster analysis result of the logarithmic *Ward*' is marked by ellipses (*lines* and *dashed lines*)

2.2.1 The Example in Detail

In Table 1, the data counts the world's largest merchant fleets by country of owner, i.e. all self-propelled oceangoing vessels 1,000 gross tons and greater (as of July 1, 2003, published by CIA World Factbook 2003). It consists of 20 observations and the variables Full Container (abbr.: Cont), Dry Bulk (Bulk), and Tanker (Tank). Hierarchical clustering gives a single unique solution, a hierarchy of nested partitions. This is in opposition to the partitional K-means clustering that leads to locally optimal solutions depending on initial partitions. Figure 1 shows the result of hierarchical clustering by *Ward*'s method.

In Fig. 2, the four cluster solutions of both the hierarchical *Ward*'s and the logarithmic *Ward*'s method are shown in the projection space that is obtained by correspondence analysis (CA). The latter is an appropriate visualisation tool for such data and for clusters. Moreover, the CA plots will become more informative by an additional projection of a dendrogram as shown in Fig. 3. In Fig. 2, the

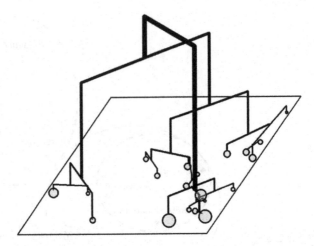

Fig. 3 Bubble-plot-dendrogram coming from the logarithmic *Ward*'s method. The size of the bubble of the corresponding observation is proportional to its mass. The latter quantifies the importance of the observations in the statistical analysis

CA plot reflects the true chi-square distances completely, i.e., to 100 %. So, it is not only an approximation as usual in the case of high dimensionality. The row points (countries) are marked by diamonds and the column points by squares. The criterion value (sum of within-cluster inertia) equals 0.04319 (Ward) and 0.04148 (logarithmic Ward), respectively. The latter classifies Malaysia and Korea in the cluster {Denmark, Singapore, United Kingdom} that was found by *Ward*'s method. By the way, the four cluster solution of *Ward*'s method can be improved by the partitional $D_{ih}Ex$ clustering to 0.04209 by shifting Malaysia to the cluster {Denmark, Singapore, United Kingdom}. So, in this application, both methods the SS and the logarithmic SS come to similar results.

In addition to Fig. 2, Fig. 3 shows the dendrogram of the hierarchical logarithmic *Ward*'s method that is projected onto the plan of CA. Such a bubble-plot-dendrogram is important for a better understanding of the results of both CA and clustering because the size of a bubble is proportional to the importance of the corresponding observation. For example, the compact cluster at the lower right corner (consisting of Greece, Japan, China,..., see also Fig. 2) contains more than half of all masses. It is merged at the end.

3   Built-in Validation

In hierarchical clustering the pivotal question arises: How to choose the number of clusters (i.e., the distance levels) at which the dendrograms should be cut in Fig. 1? Simulation studies based on bootstrapping can help to answer this question.

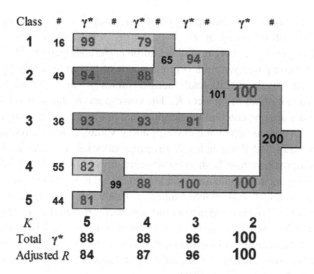

Fig. 4 Informative dendrogram of the Swiss bank data based the results of built-in validation (the cardinality of clusters is given below the symbol #). Here the Jaccard measure γ^* (values are in %) evaluates the stability of every cluster. The total Jaccard index and the adjusted Rand index R at the *bottom* (also in %) can be used to decide about the number of clusters. Obviously, the two cluster solution is most stable

Different resampling techniques can be carried out by "playing" with the weights of observations in special random ways, see for instance (Mucha 2004). The stability of results of hierarchical and partitional clustering can be assessed based on measures of correspondence between partitions and/or between clusters. Finding the appropriate number of clusters is the main task apart from individual cluster validation. One gets many bootstrap results instead of an usual unique result of hierarchical cluster analysis. The built-in validation techniques (as proposed by Mucha 2009) evaluate additionally the stability of each cluster and the degree of membership of each observation to its cluster. Using special randomized weights of objects one can easily perform built-in validations of cluster analysis results via bootstrapping techniques. There are several measures of similarity between two clusterings (partitions) (Hubert and Arabie 1985) and between sets (clusters) (Hennig 2007). The most well-known measures are the Jaccard index and the adjusted Rand index.

Figure 4 presents the simulation result of hierarchical clustering of the well-known Swiss bank data (Flury and Riedwyl 1988) by *Ward*'s method. The data consists of measurements on 200 Swiss bank notes, where 100 are genuine (i.e., real) bank notes and 100 are forged ones. The following six variables are measured: length of the bill, height of the bill on the left side, height of the bill on the right side, distance of the inner frame to the lower border, distance of the inner frame to the upper border, and length of the diagonal of the central picture. *Ward*'s method performs very well: one observation is misclassified only.

Here, for validation, soft resampling ("soft bootstrapping") is used, i.e., random weighting of the observations by $m_i = z_i * 1.8 + 0.1$, where $z_i \in [0, 1)$ is a uniform random generated weight of observation i. Here altogether 250 "bootstrap-hierarchies" are compared with the original result of hierarchical cluster analysis. The comparisons result in 250 total Jaccard values γ^* and 250 adjusted Rand values R for each number of clusters K. The corresponding Jaccard values of each individual cluster are presented in the dendrogram in Fig. 4. At the bottom, the total Jaccard index (averaged across the corresponding values of the individual clusters above) and the adjusted Rand index R (average value) give quite similar results. The (theoretical) maximum of both similarity measures is 1 (or 100 %). Without any doubt, the two cluster solution is very stable with regard to both measures ($= 100\%$, or more exactly: total $\gamma^* = 0.9996816$ and $R = 0.999356$). The stability of clusters strongly depends on how homogeneous and how well separated from neighbouring clusters they are. The latter can be observed well in hierarchical cluster analysis. For example, the stable cluster of genuine bank notes in the lower part of Fig. 4 ($= 99$ observations) becomes instable for K equals 4: The Jaccard measure decreases from 100 % (for $K = 2$ and $K = 3$) to 88 %.

4   Conclusions

Hierarchical and partitional clustering based on pairwise proximities are in some sense more general. Gaussian model-based cluster analysis in its simplest settings can deal effectively with weighted observations. Using special weights of both the observations and the variables, simple model-based Gaussian cluster analysis allows the segmentation of a contingency table by minimization of the sum of within-cluster inertia. Moreover, pairwise clustering is very easy to apply. Based on CA several graphics can be recommended for a presentation of data and clusters. The stability of results of pairwise clustering can be assessed by bootstrap techniques based on random weights of the objects. This can be done very effectively because the pairwise distances are not affected by random weighting.

References

Banfield JD, Raftery AE (1993) Model-based Gaussian and non-Gaussian clustering. Biometrics 49(3):803–821
Bartel HG, Mucha HJ, Dolata J (2003) Über eine Modifikation eines graphentheoretisch basierten partitionierenden Verfahrens der Clusteranalyse. Match 48:209–223
CIA World Factbook (2003) World's largest merchant fleets by country of owner. http://www.geographic.org
Flury B, Riedwyl H (1988) Multivariate statistics: a practical approach. Chapman and Hall, London
Greenacre MJ (1988) Clustering the rows and columns of a contingency table. J Classif 5:39–51

Hennig C (2007) Cluster-wise assessment of cluster stability. Comput Stat Data Anal 52(1): 258–271

Hubert L, Arabie P (1985) Comparing partitions. J Classif 2:193–218

Mucha HJ (2004) Automatic validation of hierarchical clustering. In: Antoch J (ed) Proceedings in computational statistics, COMPSTAT 2004, Prague. Physica-Verlag, Heidelberg, pp 1535–1542

Mucha HJ (2009) ClusCorr98 for Excel 2007: clustering, multivariate visualization, and validation. In: Mucha HJ, Ritter G (eds) Classification and clustering: models, software and applications. Report 26, WIAS, Berlin, pp 14–40

Mucha HJ, Bartel HG, Dolata J (2005) Techniques of rearrangements in binary trees (dendrograms) and applications. Match 54:561–582

Späth H (1985) Cluster dissection and analysis. Horwood, Chichester

Classification, Clustering, and Visualisation Based on Dual Scaling

Hans-Joachim Mucha

Abstract In practice, the statistician is often faced with data already available. In addition, there are often mixed data. The statistician must now try to gain optimal statistical conclusions with the most sophisticated methods. But, are the variables scaled optimally? And, what about missing data? Without loss of generality here we restrict to binary classification/clustering. A very simple but general approach is outlined that is applicable to such data for both classification and clustering, based on data preparation (i.e., a down-grading step such as binning for each quantitative variable) followed by dual scaling (the up-grading step: scoring). As a byproduct, the quantitative scores can be used for multivariate visualisation of both data and classes/clusters. For illustrative purposes, a real data application to optical character recognition (OCR) is considered throughout the paper. Moreover, the proposed approach will be compared with other multivariate methods such as the simple Bayesian classifier.

1 Introduction

We consider here binary classification and clustering based on the dual scaling technique. However, our approach of binary classification is not restricted explicitly to $K = 2$ classes. For $K > 2$ classes, this results in $(K - 1)K/2$ binary classifiers. This is denoted as pairwise classification because one has to train a classifier for each pair of the K classes.

The dual scaling classification/clustering (DSC) technique proposed here is motivated by practical problems of analyzing a huge amount of mixed data efficiently. (Concerning an introductory data mining textbook, see Berry and Browne 2006.)

H.-J. Mucha (✉)
Weierstrass Institute for Applied Analysis and Stochastics (WIAS), D-10117 Berlin, Germany
e-mail: mucha@wias-berlin.de

W. Gaul et al. (eds.), *German-Japanese Interchange of Data Analysis Results*,
Studies in Classification, Data Analysis, and Knowledge Organization,
DOI 10.1007/978-3-319-01264-3_5,
© Springer International Publishing Switzerland 2014

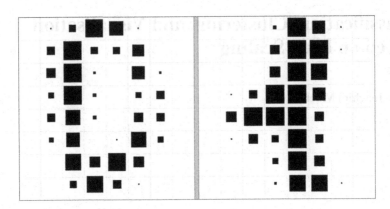

Fig. 1 Examples of images of the digits 0 and 1 in a 8 × 8 grid of pixels. The square of a pixel represents an integer value ranging from 0 (empty square) to 16 (full square)

One way to deal with such problems is down-grading all data to the lowest scale level, that is, downgrading to categories by loosing all quantitative information. Another general way is binary coding (Kauderer and Mucha 1998) which is much more expensive in computer space and time.

Throughout the paper, an application of the proposed methods to OCR is presented, see on the UCI – website http://archive.ics.uci.edu/ml/datasets/Optical+ Recognition+of+Handwritten+Digits (Frank and Asuncion 2010; Vamvakas et al. 2010). (By the way, the paper of Parvez and Mahmoud (2013) presents a recent state of the art OCR application.) The main focus of this paper is on classification. However it will be shown that DSC can be used in an iterative manner also for clustering. The basis of the ORC data are normalized bitmaps of handwritten digits from a preprinted form. From a total of 43 people, 30 contributed to the training set and the other 13 to the test set. The resulting 32 × 32 bitmaps are divided into non-overlapping blocks of 4 × 4 and the number of pixels are counted in each block. This generates an input matrix of 8 × 8 where each element is an integer in the range 0 . . . 16. This reduces the dimensionality and gives invariance to small distortions. Figure 1 shows examples of digits 0 and 1: the first observation of "0" and "1" of the database, respectively. The area of a square is proportional to the integer value of the corresponding element of the 8 × 8 matrix (64 variables/attributes).

Usually, these 64 ordinal variables with 17 categories at most can be directly processed by DSC. However, the number of categories is high with respect to the sample size of a few hundred and there still is an ordinal scale. Therefore (and because there is a general need of categorization in case of metric variables) here the original ORC data is down-graded exemplarily in a quite rough manner into at most five categories: "never" (count = 0), "seldom" (1–4), "sometimes" (5–11), "often" (12–15), and "always" (16). The corresponding transformation to do this is simply: if($c = 0$;"never";if($c = 16$;"always";if($c > 11$;"often";if($c < 5$;"seldom"; "sometimes")))) (based on the standard function if(test;then;else), and where c

never	seldom	sometimes	often	often	seldom	never	never		1	2	3	4	5	6	7	8
never	sometimes	always	sometimes	sometimes	sometimes	never	never		9	10	11	12	13	14	15	16
never	sometimes	always	seldom	never	sometimes	seldom	never		17	18	19	20	21	22	23	24
never	sometimes	always	seldom	never	sometimes	sometimes	never		25	26	27	28	29	30	31	32
never	sometimes	often	seldom	never	sometimes	sometimes	never		33	34	35	36	37	38	39	40
never	seldom	often	never	seldom	often	sometimes	never		41	42	43	44	45	46	47	48
never	never	often	sometimes	often	sometimes	never	never		49	50	51	52	53	54	55	56
never	never	sometimes	often	sometimes	seldom	never	never		57	58	59	60	61	62	63	64

Fig. 2 The values of the observation from the class Digit "0" of Fig. 1 after transformation into text data (on the left hand side) and numbering of the 64 variables

stands for count). Figure 2 (on the left hand side) shows a result of binning into categorical data as described by the simple formula above. In addition, on the right hand side, the ordering of the 64 variables in a row vector of the data matrix $\mathbf{X}$ (see below) is given. The quantitative meaning of the target words is only for illustrative purposes. DSC makes no use of any ordering. That means that almost all quantitative information in the data is lost beforehand. However, especially in medical applications, special optimal scaling techniques for ordered categories can be of great interest (Pölz 1995, 1996).

2 Binary Classification Based on Dual Scaling

The (actual) starting point of DSC is a $(I \times J)$-data table $\mathbf{X} = (x_{ij})$, $i = 1, 2, \ldots, I$, $j = 1, 2, \ldots, J$ of I observations and J variables. It is only supposed for DSC that the number of different values (or categories, words, ...) of each variable should be at least two. (Variables consisting of only one value do not effect the result.) On the other hand, for reasons of getting a stable result for classifying unseen observations, the number of categories should be "as small as possible" with regard to the number of observations. Figure 2 shows an example of an observation. In addition to $\mathbf{X}$, a binary class membership variable is required. It will be used in order to give categorical data a quantitative meaning. Generally, we want to obtain a new variable $\mathbf{z}_j$ so as to make the derived scores within the given classes ss_w as similar as possible and the scores between classes ss_b as different as possible (Nishisato 1980, 1994). The basis is a contingency table, which can be obtained by crossing a categorical variable $\mathbf{x}_j$ of M_j categories with the class membership variable. That is, regarding some constraints in the frame of a dual scaling approach, the squared correlation ratio has to be maximized:

$$\eta^2 = \frac{ss_b}{ss_b + ss_w} \quad \left(= \frac{ss_b}{ss_t}\right). \tag{1}$$

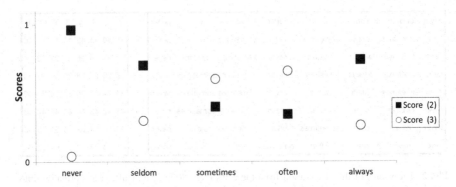

Fig. 3 Plot of the scores (2) and (3) of variable 14, respectively

Because of the well-known variance decomposition

$$ss_t = ss_w + ss_b,$$

the squared correlation ratio (1) lies in the interval [0,1].

Considering the special case of two classes, dual scaling can be applied without the calculation of orthogonal eigenvectors. Without loss of generality, a given category y_{mj} $(m = 1, 2, \ldots, M_j)$ of a variable j is transformed into an optimally scaled variable in the sense of maximal between classes variance by

$$u_{mj} = \frac{p_{mj}^{(1)}}{p_{mj}^{(0)} + p_{mj}^{(1)}}, \quad j = 1, 2, \ldots, J, m = 1, 2, \ldots, M_j. \tag{2}$$

Here $p_{mj}^{(0)}$ is an estimate of the probability for being a member of class 0 when coming from category m of variable j, whereas on the other side $p_{mj}^{(1)}$ is an estimate of the probability for being a member of class 1 when coming from category m of variable j. Alternatively, in the case of two classes, one can use the scoring

$$v_{mj} = \frac{p_{mj}^{(0)}}{p_{mj}^{(0)} + p_{mj}^{(1)}}, \quad j = 1, 2, \ldots, J, m = 1, 2, \ldots, M_j. \tag{3}$$

because of $v_{mj} = 1 - u_{mj}$. Therefore, without loss of generality, the scoring (2) will be considered here later on. Figure 3 shows an example of the estimated probabilities (scores) (2) and (3). They are estimated simply based on contingency tables. Table 1 shows both a contingency table and the estimated scores (2) of variable 14.

The final result of the transformation (2) is a quantitative data matrix $\mathbf{Z} = (z_{ij})$, $i = 1, 2, \ldots, I, j = 1, 2, \ldots, J$. Figure 4 shows a special heatmap of the distance matrix of 14 observations (selected from 765 observations in total) based on $\mathbf{Z}$. The smaller the square in a cell of the 14×14 grid the lower is the corresponding distance. Here the squared euclidean distance is applied. The 6×6 square area

Table 1 Contingency table as the result of cross-tabulation of variable 14 and the class membership variable, and estimated scores (2)

Category	Given class Digit "0"	Digit "1"	Total	Score
Never	2	49	51	0.9608
Seldom	23	54	77	0.7013
Sometimes	171	114	285	0.4000
Often	146	76	222	0.3423
Always	34	96	130	0.7385
Total	376	389	765	0.5085

on the left top side presents the distances between observations of the class Digit "0". Obviously, there are small within-class distances and this class looks very homogeneous. In opposition, the many different sizes of the squares on the right bottom side represent distances between observations of the class Digit "1". This class looks inhomogeneous. All other squares are big. They indicate great distances between observations from different classes. In Fig. 5, the principal component analysis (PCA) plot shows a multivariate view of $\mathbf{Z}$. Obviously, it emphasizes the inhomogeneity of the class Digit "1". For further visualization methods, see Mucha (2009).

On the other side the solution (score value) f_{mj} of the well-known correspondence analysis (Greenacre 1984) that depends on the actual sample size of the two clusters is simply related to (2) by

$$u_{mj} = \frac{\sqrt{I_0 I_1}}{(I_0 + I_1)} f_{mj} + c, \quad j = 1, 2, \ldots, J, m = 1, 2, \ldots, M_j,.$$

where I_0 and I_1 is the number of observations in cluster 0 and cluster 1, respectively. The constant c is responsible for the shift of the mean value of the scores f_{kj} to the origin. That is, every variable j $(j = 1, 2, \ldots, J)$ is scaled by the same value $b = \sqrt{I_0 I_1}/I$ and shifted by a constant c.

2.1 Distance Scores and the Cut-off Value

After upgrading the categories all data values of $\mathbf{Z} = (z_{ij})$ are in the interval $[0, 1]$. Without loss of generality the hypothetical worst-case $\mathbf{z}_w \equiv \mathbf{0}$ is considered here (this naming comes from credit scoring where class 0 stands for bad applicants, see Mucha et al. 1998). (Otherwise one can look at the best case model with $\mathbf{z}_w \equiv \mathbf{1}$.) Then the Manhattan distance t_{iw} between an observation $\mathbf{z}_i$ and the worst case $\mathbf{z}_w$ has both the suitable and really simple form

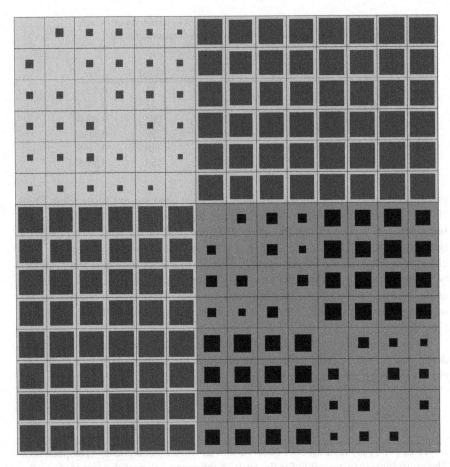

Fig. 4 Class-heatmap of a 14×14 distance matrix that is figured out by using the scores from dual scaling. The first six observations come from class Digit "0" (light gray background color) and the last eight from class Digit "1" (gray background color). As before in Fig. 1, the size of a square is proportional to the corresponding distance value

$$t_{iw} = t(\mathbf{z}_i, \mathbf{z}_w) = \sum_{j=1}^{J} z_{ij}. \tag{4}$$

We call t_{iw} the (multivariate) distance score. Figure 6 visualizes the relation between the distance scores and the error rate with respect to the class membership variable. Now the question is how to build up a classifier based on the distance scores. One simple way is looking for a suitable cut-off-point on the distance score axis. Obviously, in Fig. 6, the optimum error rate is near 0 (only one error occurs, marked by a great circle, with respect to 765 observations). There is a wide range of this low error rate on the distance axis, see the marking below the abscissa. In Fig. 6,

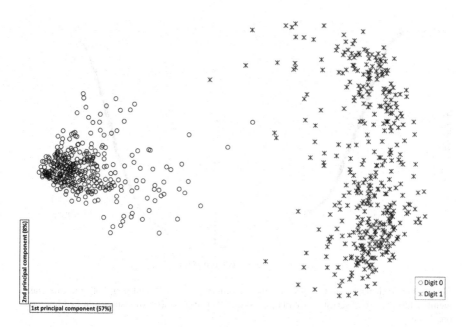

Fig. 5 PCA plot of the two classes Digit "0" and Digit "1" after the dual scaling (upgrading) step. In the training data set, the cardinalities of the classes are 376 and 389, respectively

the dashed line highlights an appropriate cut-off-threshold. Here the corresponding cut-off-value of the minimum error rate is 0.527. It is simply the average of the two scores of minimum error rate. This cut-off-value is used as a classifier later on for decision-making on new observations. Obviously, the two classes seem to be very well separated based on the training data set.

The dual scaling classification of the observations of the test data set from the UCI database results in an error rate of 0.28 % with respect to the cut-off-value estimated from the training data set given above. Figure 7 shows the relation between the distance scores (4) of the test data and the error rate. One error is counted only. It is marked by a great circle. In addition, the distance scores of the training data are drawn. Both curves of scores are very close to each other. For numerical details concerning the error rates see Tables 2 and 3. Mucha et al. (1998) investigated the stability of a special local DSC approach in an application to credit scoring. It was the most stable classification method in comparison with five other multivariate methods such as neural networks and C4.5.

2.2   The Naive Bayes Classifier

The simple Bayesian classifier (SBC), sometimes called Naive-Bayes, is based on a conditional independence model of each variable given the class. Once the

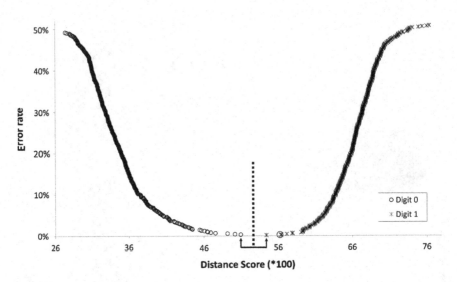

Fig. 6 Plot of distance scores t_{iw} versus the classification error. In addition, both the true class membership of each observation (marked by symbols) and the cut-off-threshold (dashed line) are presented

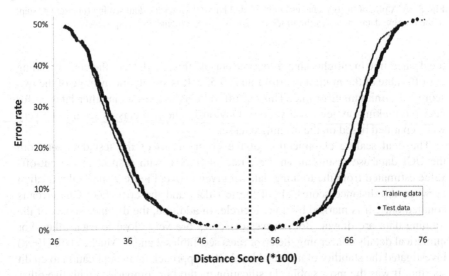

Fig. 7 Plot of distance scores t_{iw} of the test data versus the classification error. In addition, both the distance scores of the training data (see Fig. 6) and the cut-off-threshold are presented

discrete probabilities $p_{mj}^{(0)}$ and $p_{mj}^{(1)}$ are estimated in DSC (see above and Table 1), the posterior probabilities can be computed easily as a by-product. In the ORC two class data, categorised as shown in Fig. 2, the SBC performs similar to DSC: also one observation (but not the same one) is misclassified. However, the error rate of

Table 2 Training data set: comparison of results of dual scaling and Naive Bayes classification

	Dual scaling classification			Naive Bayes classification		
	Confusion matrix			Confusion matrix		
Given classes	Class 0	Class 1	Error rate (%)	Class 0	Class 1	Error rate (%)
Digit "0"	375	1	0.27	375	1	0.27
Digit "1"	0	389	0.00	0	389	0.00
Total	376	389	0.13	375	390	0.13

Table 3 Test data set: comparison of results of dual scaling and Naive Bayes classification

	Dual scaling classification			Naive Bayes classification		
	Confusion matrix			Confusion matrix		
Given classes	Class 0	Class 1	Error rate (%)	Class 0	Class 1	Error rate (%)
Digit "0"	177	1	0.56	176	2	1.12
Digit "1"	0	182	0.00	2	180	1.10
Total	178	183	**0.28**	178	182	**1.11**

SBC of the test data is a little bit higher: four observations are misclassified. Tables 2 and 3 summarize the comparison of the two methods using the training data set and the test data set, respectively.

Similar good results can be reported also from linear discriminant analysis, where, of course, the original data set is used.

2.3 Dual Scaling Cluster Analysis

Here the starting point is a random generated class membership variable $P^0(I, 2)$ of the I observations into 2 clusters. The iterative DSC repeats both the upgrading step and the clustering step until either no change of two successive partitions $P^{n+1}(I, 2)$ and $P^n(I, 2)$ occurs or until a fixed number of iterations are done. For further details see Mucha (2002). Generally speaking the iterative DSC looks for directions in the space and changes the geometry of the points (observations) more in a random way than in a direct way of optimization. This is done partly by optimum univariate optimization on the one hand and on the other hand by some heuristics. Moreover, a result is obtained in any case independent of the existence of a class structure. Here the voting technique is beside univariate dual scaling another important key for clustering by DSC. Voting is an ensemble technique. Voting means both finding the most typical partition (MTP) among a set of partitions that are obtained simultaneously by starting from different initial partitions (Mucha 2002). Obviously there are several ways in defining a MTP. Here, in the two class setting, a MTP is chosen to be the one which sum over all simple matching similarities s_{mn} to the remaining partitions becomes a maximum. Taking into account the random permutation of cluster numbering the simple matching coefficient for two partitions $P^m(I, 2)$ and $P^n(I, 2)$ is defined as

$$s_{mn} = Max(\frac{a+d}{I}, \frac{b+c}{I}),$$

where $(a + d)$ count the number of matches and $(b + c)$ count the number of mismatches, respectively. In the ORC two class data, convergence is reached usually after seven iterations depending on the start partition. Herein the procedure of voting is based on ten partitions. In that way, the iterative DSC finds both quite similar quantifications and distance scores as the corresponding classification method described before.

3 Conclusion

The idea of dual scaling classification is presented here in a detailed way. DSC is both the most simple and computationally efficient classification method. Obviously, it seems to be stable against outliers in the data. Missing values are allowed simply by an additional category. DSC changes the geometry of the data points in the case of given quantitative variables. On the basis of this new geometry one can apply multivariate projection methods such as principal components analysis in order to make a previously invisible structure visible. Sure, DSC can be improved by sophisticated categorization methods.

References

Berry MW, Browne M (eds) (2006) Lecture notes in data mining. World Scientific, Singapore

Frank A, Asuncion A (2010) UCI machine learning repository. School of Information and Computer Sciences, University of California, Irvine. http://archive.ics.uci.edu/ml

Greenacre MJ (1984) Theory and applications of correspondence analysis. Academic, London

Kauderer H, Mucha HJ (1998) Supervised learning with qualitative and mixed attributes. In: Balderjahn I, Mathar R, Schader M (eds) Classification, data analysis, and data highways. Springer, Berlin, pp 374–382

Mucha HJ (2002) An intelligent clustering technique based on dual scaling. In: Nishisato S, Baba Y, Bozdogan H, Kanefuji K (eds) Measurement and multivariate analysis. Springer, Tokyo, pp 37–46

Mucha HJ (2009) ClusCorr98 for Excel 2007: clustering, multivariate visualization, and validation. In: Mucha HJ, Ritter G (eds) Classification and clustering: models, software and applications. Report 26, WIAS, Berlin, pp 14–40

Mucha HJ, Siegmund-Schultze R, Dübon K (1998) Adaptive cluster analysis techniques – software and applications. In: Hayashi C, Ohsumi N, Yajima K, Tanaka Y, Bock HH, Baba Y (eds) Data science, classification and related methods. Springer, Tokyo, pp 231–238

Nishisato S (1980) Analysis of categorical data: dual scaling and its applications. University of Toronto Press, Toronto

Nishisato S (1994) Elements of dual scaling: an introduction to practical data analysis. Lawrence Erlbaum Associates, Hillsdale

Parvez MT, Mahmoud SA (2013) Arabic handwriting recognition using structural and syntactic pattern attributes. Pattern Recognit 46(1):141–154

Pölz W (1995) Optimal scaling for ordered categories. Comput Stat 10:37–41

Pölz W (1996) Überprüfung und Erhöhung der Diskriminierfähigkeit von Skalen. In: Mucha HJ, Bock HH (eds) Classification and multivariate graphics: models, software and applications. Report 10, WIAS, Berlin, pp 51–55

Vamvakas G, Gatos B, Perantonis SJ (2010) Handwritten character recognition through two-stage foreground sub-sampling. Pattern Recognit 43(8):2807–2816

Variable Selection in K-Means Clustering via Regularization

Yutaka Nishida

Abstract In many cases, both essential and irrelevant variables to the cluster structure are included in the data set. The K-means algorithm (MacQueen JB (1967) Some methods for classification and analysis of multivariate observations. In: Proceedings of the 5th Berkeley symposium on mathematical statistics and probability, Berkeley, pp 281–297) which is one of the most popular clustering method can treat such a data set, but can not identify the essential variables for clustering. In supervised-learning methods such as regression analysis, variable selection is a major topic. However, variable selection in clustering currently is not an active area of research. In this study, a new method of K-means clustering is proposed to detect irrelevant variables to the cluster structure. The proposed method achieves the purpose of calculating variable weights using an entropy regularization method (Miyamoto S, Mukaidono M (1997) Fuzzy c-means as a regularization and maximum entropy approach. In: Proceedings of the 7th international Fuzzy Systems Association World Congress, Prague, vol 2, pp 86–92) which is developed to obtain fuzzy memberships in fuzzy clustering. This method allows us to identify the important variables for clustering.

1 Introduction

In many cases, both essential and irrelevant variables to the cluster structure are included in the data set. The K-means algorithm (MacQueen 1967) which is one of the most popular clustering methods can treat such data sets, but can not distinguish which variable is essential to the cluster structure.

Y. Nishida (✉)

Graduate School of Human Sciences, Osaka University, 1-2 Yamadaoka, Suita, Osaka, Japan
e-mail: nishida@bm.hus.osaka-u.ac.jp

W. Gaul et al. (eds.), *German-Japanese Interchange of Data Analysis Results*, 71
Studies in Classification, Data Analysis, and Knowledge Organization,
DOI 10.1007/978-3-319-01264-3_6,

There are two approaches for fixing this problem. One is the extraction of the cluster structure by using a dimension reduction method such as principal component analysis. Since this approach needs an interpretation of the components, the interpretation of a result needs expert domain knowledge and is not easy. The other approach is variable selection. In this approach, the importance of variables is represented as a vector of variable weights, so a user can interpret the result easily.

In regression analysis, discriminant analysis and machine learning (e.g. with support vector machines), variable selection is a major topic. However, variable selection in clustering is not an area of active research. In this study, a new method of K-means clustering is proposed to detect irrelevant variables to the cluster structure.

Let $\mathbf{X} = \{x_{ij}\}$ be an $N \times P$ data matrix, where x_{ij} is an element of the i-th object $(i = 1, \ldots, N)$ in the j-th variable $(j = 1, \ldots, P)$, $\mathbf{U} = \{u_{ik}\}$ be a membership matrix, where u_{ik} is a membership of the i-th object which belongs to cluster k with $k = 1, \ldots, K$, and $\bar{\mathbf{X}} = \{\bar{x}_{kj}\}$ be the $K \times P$ centroid matrix. The objective of the K-means algorithm is to minimize the within-cluster sum of squares:

$$F(\mathbf{U}, \bar{\mathbf{X}}) = \sum_{k=1}^{K} \sum_{i=1}^{N} \sum_{j=1}^{P} u_{ik} d_{ikj}, \tag{1}$$

where $d_{ikj} = (x_{ij} - \bar{x}_{kj})^2$. Though the K-means algorithm treats all variables in the same way when the distance between cluster centers and data points is calculated, a certain variable might be more important than other variables. In this study, the weights for variables are introduced into the distance as

$$d_{ikj} = w_j (x_{ij} - \bar{x}_{kj})^2, \tag{2}$$

where w_j satisfies the constraints $\sum_{j=1}^{P} w_j = 1$ and $0 \leq w_j \leq 1$.

By introducing the weight parameter, we can identify the irrelevant variables to the cluster structure. However, the solution obtained by optimizing the objective function with weights parameter $\mathbf{w}$ is not a desirable for variable selection. The solution is given by $w_j = 1$ or 0. This means only one variable is always selected.

Both the objective function of K-means with weights parameter $\mathbf{w}$ and it's constraints ($\sum_{j=1}^{P} w_j = 1$ and $0 \leq w_j \leq 1$) are linear with respect to the introducing parameter w_j. This problem is a linear programing and it's optimal solution is at a vertex of the hyper-polygon of the feasible solutions is applicable. In this case, we have $w_j = 1$ or 0. This property implies that it is necessary to introduce the nonlinearity into the objective function for $\mathbf{w}$ to obtain fuzzy weight. This is the same problem of the estimating fuzzy u_{ik} in fuzzy c-means (see Miyamoto et al. 2008). Therefore, in order to estimate $\mathbf{w}$, we can utilize the same techniques as in the estimation of fuzzy u_{ik} in fuzzy c-means algorithm.

In this paper, the entropy regularization method (Miyamoto and Mukaidono 1997) is utilized to estimate the weights. The entropy regularization method is developed to obtain fuzzy memberships parameter which satisfy the conditions $\sum_{i=1}^{N} u_{ik} = 1$ and $0 \leq u_{ik} \leq 1$ in fuzzy clustering.

In the following, we use this approach for variable selection. By using the entropy regularization method, it is possible to estimate the parameter w_j that satisfies the constraints. The values of parameter $\mathbf{w}$ being close to zero means corresponding variables are irrelevant for clustering. We consider variable weights take the same value each variables in ordinary K-means algorithm. Our approach extend this fixed weight parameter to take fuzzy value, $[0, 1]$ using entropy regularization method. Huang et al. (2005) took a similar approach and proposed W-K-means algorithm which utilizes the fuzzy c-means algorithm (Bezdek 1980). The objective function of W-K-means is written as

$$F(\mathbf{U}, \bar{\mathbf{X}}, \mathbf{w}) = \sum_{k=1}^{K} \sum_{i=1}^{N} \sum_{j=1}^{P} u_{ik} w_j^{\beta} (x_{ij} - \bar{x}_{kj})^2, \tag{3}$$

where β ($\beta < 0$ or $\beta > 1$) is a parameter for weight w_j.

In the Sect. 2, we show the objective function of a new K-means method with variable selection and the parameter estimation algorithm for the proposed method. In the Sects. 3 and 4, the usefulness of the proposed method is demonstrated with synthetic and real data. Finally, we present conclusions in the Sect. 5.

2 Method

2.1 Objective Function

The objective function of a new K-means method can be defined as

$$F(\mathbf{U}, \bar{\mathbf{X}}, \mathbf{w}) = \sum_{k=1}^{K} \sum_{i=1}^{N} \sum_{j=1}^{P} u_{ik} w_j (x_{ij} - \bar{x}_{kj})^2 + \lambda^{-1} \sum_{j=1}^{P} w_j \log w_j, \tag{4}$$

where $\mathbf{w}$ is a $P \times 1$ variable weight vector consisting of w_j as an element, and $\lambda > 0$ is a regularizing penalty parameter for variable weights. The first term is the within-cluster sum of squares and the second term is the regularizing penalty term for variable weights. In this objective function, the entropy regularization method (Miyamoto and Mukaidono 1997) is utilized to achieve the purpose of calculating variable weights.

The membership parameter $\mathbf{U}$ and the weight parameter $\mathbf{w}$ satisfy the following conditions.

$$\sum_{k=1}^{K} u_{ik} = 1, \ u_{ik} \in \{0, 1\}, \tag{5}$$

$$\sum_{j=1}^{P} w_j = 1, \ 0 \leq w_j \leq 1. \tag{6}$$

The constraint (5) implies that an object belongs to one and only one cluster and (6) means the total of the weights is 1. $\mathbf{w}$ is the weight vector of the variables.

2.2 Algorithm

To estimate the parameters $(\mathbf{U}, \bar{\mathbf{X}}, \mathbf{w})$, an iterative algorithm is developed to minimize the objective function (4), given the value of λ. Specifically, this algorithm has the following three steps.

- U step: Update $\mathbf{U}$ for fixed $\bar{\mathbf{X}} = \bar{\mathbf{X}}^*$, $\mathbf{w} = \mathbf{w}^*$.

  This step is very similar to the ordinary K-means algorithm except for the parameter w_j.

$$u_{ik} = \begin{cases} 1, \ \text{if} \displaystyle\sum_{j=1}^{P} w_j (x_{ij} - \bar{x}_{kj})^2 \leq \sum_{j=1}^{P} w_j (x_{ij} - \bar{x}_{lj})^2, \\ 0, \ \text{otherwise.} \end{cases} \tag{7}$$

- $\bar{\mathbf{X}}$ step: Update $\bar{\mathbf{X}}$ for fixed $\mathbf{U} = \mathbf{U}^*$, $\mathbf{w} = \mathbf{w}^*$.

  This step is the same as for the ordinary K-means algorithm:

$$\bar{x}_{kj} = \frac{\displaystyle\sum_{i=1}^{N} u_{ik} x_{ij}}{\displaystyle\sum_{i=1}^{N} u_{ik}}. \tag{8}$$

- w step: Update $\mathbf{w}$ for fixed $\mathbf{U} = \mathbf{U}^*$, $\bar{\mathbf{X}} = \bar{\mathbf{X}}^*$.

  This optimal solution is derived from solving the constrained optimization problem with the method of Lagrange multipliers (see Appendix).

$$w_j = \frac{\exp(-\lambda \displaystyle\sum_{k}^{K} \sum_{i}^{N} u_{ik}(x_{ij} - \bar{x}_{kj})^2)}{\displaystyle\sum_{m=1}^{P} \exp(-\lambda \displaystyle\sum_{k}^{K} \sum_{i}^{N} u_{ik}(x_{im} - \bar{x}_{km})^2)}. \tag{9}$$

3 Simulation Study

In this section, we conduct a simulation study to evaluate whether the proposed method can estimate the true cluster structure and the frequency of local minima. In the beginning, we describe the procedure of generating data sets. The data set is assumed to be composed of the part with the cluster structure and the part that doesn't have it. Let $\mathbf{X}^{(t)}$ be an $N \times p$ structured data matrix and $\mathbf{X}^{(e)}$ be an $N \times q$ noise part data matrix, where p is the number of relevant and q is the number of irrelevant variables. An $N \times (p + q)$ data matrix is written as $\mathbf{X} = [\mathbf{X}^{(t)}, \mathbf{X}^{(e)}]$.

$\mathbf{X}^{(t)}$ is generated from multivariate normal distributions with unique mean vectors in each cluster and with the same variance between clusters and variables (Adachi 2011). A variable vector of l-th object which belongs to cluster k, $\mathbf{x}_{l(k)}^{(t)}$ is generated from the p-variate normal distribution $N_p(\boldsymbol{\mu}_k, v(\theta)\mathbf{I}_p)$ with mean vector $\boldsymbol{\mu}_k$ and covariance matrix $v(\theta)\mathbf{I}_p$, where $\mathbf{I}_p$ is an identity matrix. The variance $v(\theta)$ is a function of a degree of separation between clusters θ and defined as

$$v(\theta) = \frac{SSB \times (1 - \theta)}{N\theta},$$

where $SSB = \sum_{k=1}^{K} N_k(\boldsymbol{\mu}_k - \bar{\boldsymbol{\mu}})^2$ is between-cluster sum of squares, $\theta = SSB/(SSB+SSW) = SSB/(SSB+Nv(\theta))$ is a degree of separation between clusters, $\bar{\boldsymbol{\mu}} = \sum_{k=1}^{K} N_k\boldsymbol{\mu}_k$ is a overall mean and $SSW = \sum_{k=1}^{K} N_k v(\theta) = Nv(\theta)$ is within-cluster sum of squares.

The cluster size N_k was drawn from the discrete uniform distribution on $[30, 70]$. The mean vector $\boldsymbol{\mu}_k$ was drawn from the continuous uniform distribution on $[-10, 10]$. The degree of separation between clusters θ was set to 0.9. The variable vector $\mathbf{x}_{l(k)}^{(t)}$ was generated from $N_p(\boldsymbol{\mu}_k, v(\theta)\mathbf{I}_p)$ and arranged in the row of $\mathbf{X}^{(t)}$. The elements of $\mathbf{X}^{(e)}$ are generated from the continuous uniform distribution on $[-10, 10]$. Total number of variables $p + q$ was fixed to 10 and p took three values ($p = 8, 5, 2$). This means an operation of the number of irrelevant variables. The number of clusters K was set to 3. The regularizing parameters λ for each condition ($p = 8, 5, 2$) were set as $\lambda = 0.0001, 0.0005, 0.001$, respectively. One hundred data sets were generated for each condition.

In this simulation, 50 initial values were used in each trial and the solution which gives the minimum value of the objective function was accepted as global minima. We defined the solution with the value of the objective function which does not match with the value of the objective function corresponding to global minima as local minima.

Table 1 shows the quartile values of the Adjusted Rand Index (ARI: Hubert and Arabie 1985) and the number of local minima for each condition. The ARI is a measure of the similarity between two partitions (in this case, true partition and clustering by proposed method), which has an upper bound of one. The proposed method keeps high performance in the ARI, even when the number of irrelevant variables to the clustering structure is increased. We find that the proposed method

Table 1 The values of the ARI and the number of local minima for each condition

		$p = 8$			$p = 5$			$p = 2$		
		1st	Med	3rd	1st	Med	3rd	1st	Med	3rd
ARI	Proposed	0.93	0.96	0.98	0.90	0.92	0.96	0.56	0.79	0.93
	K-means	0.92	0.96	0.98	0.60	0.91	0.94	0.02	0.33	0.46
Local minima	Proposed	9	25	37	25.8	37.5	45	38	45	48
	K-means	1	4.5	10	3	22.5	41	40.8	46	48

have many local minima, but the proposed method has high values of the ARI than ordinary K-means. Therefore, we can conclude that the algorithm is working properly and proposed method is useful in spite of many the local minima.

4 Real Data Examples

In this section, the utility of the proposed method is examined through applications to real data sets.

4.1 Iris Data Set

First, we use the iris data set (Fisher 1936) to compare our approach with the W-K-means method of Huang et al. (2005). The tuning parameter λ in the proposed method and β in the W-K-means was set as $\lambda = 0.2,\ 0.1,\ 0.05$ and $\beta = 2,\ 3,\ 4$ respectively. Table 2 shows the ARI values and the weight values for each variable. In this table, w1–w4 means weight values corresponding to variables (sepal length, sepal width, petal length, petal width).

The weights of the K-means algorithm are regarded as fix to 1/[number of variables]. So in this case, every weight equals 1/4. The proposed method and the W-K-means have almost the same value of the ARI and better than the ordinary K-means algorithm. We find the weights of the proposed method are more sparse than the W-K-means. This property is desirable for selecting variables, because it is easy to distinguish unnecessary variables.

4.2 Animal Data Set

In this subsection, we used the animal feature data set (Toyoda 2008) to illustrate the behavior of the proposed method. This data set contains 13 features (Table 3) of 30 animals (Table 4) which are represented as dummy binary variables. The proposed

Table 2 The ARI values and the weights for each variable

| | Proposed | | | W-K-means | | | |
	$\lambda = 0.2$	$\lambda = 0.1$	$\lambda = 0.05$	$\beta = 2$	$\beta = 3$	$\beta = 4$	K-means
ARI	0.89	0.89	0.89	0.87	0.89	0.89	0.73
w1	0.001	0.030	0.103	0.082	0.158	0.188	0.250
w2	0.094	0.226	0.275	0.191	0.238	0.247	0.250
w3	0.005	0.064	0.160	0.100	0.178	0.203	0.250
w4	0.900	0.680	0.462	0.627	0.425	0.362	0.250

Table 3 The estimated weights for each variable

Small	Medium	Large	2-footed	4-footed	Hair	Hoof	Mane	Feather	Hunt	Run	Fly	Swim
0.01	0.00	0.00	0.10	0.24	0.10	0.24	0.01	0.24	0.00	0.00	0.03	0.03

Table 4 Clustering result with the number of clusters for $K = 3$

	Member
I	Boar, Cow, Horse, Deer, Zebra, Elephant, Pig
II	Dog, Wolf, Fox, Bear, Monkey, Japanese raccoon, Cheetah, Tiger, Cat, Hyena, Leopard, Lion
III	Penguin, Ostrich, Chicken, Canary, Sparrow, Hawk, Owl, Eagle, Pigeon, Duck, Goose

method was applied to the animal feature data set with the number of cluster $K = 3$ and a regularizing parameter $\lambda = 1$. The estimated weights and clustering result is shown in Tables 3 and 4.

We find that 30 animals are classified into the mammals with hooves (I), without hooves (II) and birds (III) cluster. These clusters correspond exactly with the weight values. The "4-footed", "Hoof" and "Feather" variables have a high value of weights. And these variables characterize the three animal clusters.

Some variables such as body size ("Small", "Medium", "Large"), "Mane", "Hunt" and "Run" are weighted zero or almost zero. So we can find variables with high weights which and variables without weights which are not useful to classify the 30 animals.

5 Conclusion

In this paper, a new method of K-means clustering with regularization method was proposed for variable selection. We developed an alternating least square algorithm for estimating parameters. Through the simulation study and real data examples, the advantage of the proposed method was shown. The proposed method showed better performance than other clustering methods. The weight parameters facilitate the interpretation of the result.

The proposed method have a regularization parameter λ which is determined in advance of analysis. However, there is no rational method to determine this parameter and we determined empirically in this paper. It is future work how the parameter is determined.

Appendix

We describe the derivation of (9). We can rewrite (4) as

$$F(\mathbf{U}, \bar{\mathbf{X}}, \mathbf{w}) = \sum_{j}^{P} w_j D_j + \lambda^{-1} \sum_{j}^{P} w_j \log w_j, \tag{10}$$

where $D_j = \sum_k \sum_i u_{ik}(x_{ij} - \bar{x}_{kj})^2$. Let α be a Lagrange multiplier, the Lagrangean function is written as

$$\psi(\mathbf{w}, \alpha) = \sum_{j}^{P} w_j D_j + \lambda^{-1} \sum_{j}^{P} w_j \log w_j + \alpha(\sum_{j}^{P} w_j - 1). \tag{11}$$

Differentiating partially $\psi(\mathbf{w}, \alpha)$ with respect to w_j and α, we get

$$\frac{\partial \psi(\mathbf{w}, \alpha)}{\partial w_j} = D_j + \lambda^{-1}(1 + \log w_j) + \alpha, \tag{12}$$

$$\frac{\partial \psi(\mathbf{w}, \alpha)}{\partial \alpha} = \sum_{j}^{P} w_j - 1. \tag{13}$$

Solving $\partial \psi(\mathbf{w}, \alpha)/\partial w_j = 0$ gives

$$w_j = \exp(-\lambda D_j) \times \exp(-1 - \lambda \alpha). \tag{14}$$

Solving $\partial \psi(\mathbf{w}, \alpha)/\partial \alpha = 0$ yields

$$\sum_{j}^{P} w_j = 1. \tag{15}$$

By inserting (14)–(15), we obtain

$$\exp(-1 - \lambda \alpha) = \left\{ \sum_{j}^{P} \exp(-\lambda D_j) \right\}^{-1}. \tag{16}$$

Then, inserting (16) to (14) yields

$$
w_j = \frac{\exp(-\lambda D_j)}{\sum\limits_{m}^{P} \exp(-\lambda D_m)} = \frac{\exp\left\{-\lambda \sum\limits_{k}^{K} \sum\limits_{i}^{N} u_{ik}(x_{ij} - \bar{x}_{kj})^2\right\}}{\sum\limits_{m=1}^{P} \exp\left\{-\lambda \sum\limits_{k}^{K} \sum\limits_{i}^{N} u_{ik}(x_{im} - \bar{x}_{km})^2\right\}}. \tag{17}
$$

References

Adachi K (2011) Fixed size clustering with least squares permutation. Bull Data Anal Jpn Classif Soc 1(1):11–22. (in Japanese with English abstract)

Bezdek JC (1980) A convergence theorem for fuzzy isodata clustering algorithm. IEEE Trans Pattern Anal Mach Intell 2(1):1–8

Fisher RA (1936) The use of multiple measurements in taxonomic problems. Ann Eugen 7:179–188

Huang J, Ng M, Rong H, Li Z (2005) Automated variable weighting in k-means type clustering. IEEE Trans Pattern Anal Mach Intell 27(5):657–668

Hubert L, Arabie P (1985) Comparing partitions. J Classif 2:193–218

MacQueen JB (1967) Some methods for classification and analysis of multivariate observations. In: Proceedings of the 5th Berkeley symposium on mathematical statistics and probability, Berkeley, pp 281–297

Miyamoto S, Mukaidono M (1997) Fuzzy c-means as a regularization and maximum entropy approach. In: Proceedings of the 7th international Fuzzy Systems Association World Congress, Prague, vol 2, pp 86–92

Miyamoto S, Ichihashi H, Honda K (2008) Algorithms for fuzzy clustering: methods in c-means clustering with applications. Springer, Berlin

Toyoda H (2008) Introduction to data mining (Data mining nyumon). Tokyo tosho, Tokyo. (in Japanese)

Structural Representation of Categorical Data and Cluster Analysis Through Filters

Shizuhiko Nishisato

Abstract Representation of categorical data by nominal measurement leaves the entire information intact, which is not the case with widely used numerical or pseudo-numerical representation such as Likert-type scoring. This aspect is first explained, and then we turn our attention to the analysis of nominally represented data. For the analysis of a large number of variables, one typically resorts to dimension reduction, and its necessity is often greater with categorical data than with continuous data. In spite of this, Nishisato S, Clavel JG (Behaviormetrika 57:15–32, 2010) proposed an approach which is diametrically opposite to the dimension-reduction approach, for they advocate the use of doubled hyper-space to accommodate both row variables and column variables of two-way data in a common space. The rationale of doubled space can be used to vindicate the validity of the Carroll-Green-Schaffer scaling (Carroll JD, Green PE, Schaffer CM (1986) J Mark Res 23(3):271–280). The current paper will then introduce a simple procedure for the analysis of a hyper-dimensional configuration of data, called cluster analysis through filters. A numerical example will be presented to show a clear contrast between the dimension-reduction approach and the total information analysis by cluster analysis. There is no doubt that our approach is preferred to the dimension-reduction approach on two grounds: our results are a factual summary of a multidimensional data configuration, and our procedure is simple and practical.

S. Nishisato (✉)
University of Toronto, Toronto, Canada
e-mail: shizuhiko.nishisato@utoronto.ca

W. Gaul et al. (eds.), *German-Japanese Interchange of Data Analysis Results*,
Studies in Classification, Data Analysis, and Knowledge Organization,
DOI 10.1007/978-3-319-01264-3_7,
© Springer International Publishing Switzerland 2014

1   Likert-Type Scoring and Its Limitations

When we collect multiple-choice data, we often see ordered response options (e.g., Never, Sometimes, Often, Always), and we typically use Likert-type scoring (Likert 1932), namely, 1, 2, 3 and 4 for categories Never, Sometimes, Often and Always, respectively. This scoring scheme captures linear relations between questions (items). When the method was proposed 80 years ago, researchers were mainly interested in finding a set of highly correlated items, and scores from such a set were interpreted as scale values of a unidimensional scale. The product-moment correlation is a measure of linear relations and Likert-type scoring captures linear relations. Thus, their combination offered an ideal tool for constructing unidimensional scales. However, times have changed since then, and researchers' interests are now directed also towards multidimensional scales and non-linear relations. Yet, Likert-type scoring is still as popular as it was 80 years ago.

Today the Likert-type scoring is frequently used in surveys, of which purpose is to find what data are telling us, hence no need to restrict our attention only to linear relations. Likert-type scoring employs an interval scale, but if our aim is to tap into all information in data, data must be free from metric constraints and our preferred data should be expressed in nominal measurement, the suggestion on down-grading of input measurement (Nishisato 1999). In understanding this paradoxical suggestion, it is important to distinguish between the nature of nominal measurement (i.e., the least permissible number of arithmetic operations) and the scaling of nominal measurement (i.e., modifying the least constrained numerals, thus the greatest number of possible arithmetic operations, to *upgrade* its scale level). Our task lies in the scaling. The object then is to determine category scores without metric restrictions as regression on the data, that is, those values that reproduce or approximate the data in the least-squares sense (Nishisato 2012a).

2   Complete Representation of Categorical Data

Let us consider the following multiple-choice items:

- Q1. Do you support Darwin's *Theory of Evolution*?
  1 = No; 2 = yes, but not completely; 3 = yes absolutely
- Q2. Do you support death penalty?
  1 = no; 2 = a little; 3 = strongly

Suppose that we obtained responses from 10 people. Table 1 shows responses expressed in Likert-type interval-scale format (left-hand side) and the nominal-scale format(right-hand side). The distribution of 1's in the latter format represents the entire *unconstrained* information in the current data.

In the nominal-scale representation, each subject's response to one item is (1,0,0), (0,1,0) or (0,0,1). These response patterns can be regarded as coordinates of a

Table 1 Two formats of data representation		Likert		Nominal	
	Subject	Item 1	Item 2	Item 1	Item 2
	1	3	2	001	010
	2	1	3	100	001
	3	2	2	010	010
	4	3	1	001	100
	5	2	1	010	100
	6	1	3	100	001
	7	1	2	100	010
	8	3	3	001	001
	9	1	3	100	001
	10	3	2	001	010

three-dimensional graph. The responses from ten subjects can be expressed by $(f_1, 0, 0)$, $(0, f_2, 0)$ and $(0, 0, f_3)$, where $f_1 + f_2 + f_3 = 10$. When we connect the three points, we obtain a triangle. This is a geometric form of an item, which spans 2-dimensional space as a triangle. Typically two items are not perfectly correlated, thus the corresponding two triangles span a 4-dimensional space because each 3-category item requires a 2-dimensional space. The correlation between two items then can be defined as the projection of one triangle onto the space of the other variable, which Nishisato (2006) has shown to be equivalent to Cramér's coefficient.

If the number of response categories is 4, a response to each item is one of the patterns (1,0,0,0), (0,1,0,0), (0,0,1,0) and (0,0,0,1), and these coordinates generate a pyramidal form in 3-dimensional space, or a 3-polytope; if the number is 5, the item will yield a 4-polytope; and so on. The correlation between two items of a k-polytope and an m-polytope can be calculated through the forced classification procedure (Nishisato 1984; Nishisato and Baba 1999) as shown in Nishisato (2007), or by Cramér's formula.

3 Doubled Hyper-Space for Complete Representation

From the point of view of quantification theory (e.g., dual scaling, correspondence analysis), the 10×6 response-pattern matrix of Table 1 yields four components (i.e., the rank of the table minus $1 = 5 - 1 = 4$, see Nishisato 1980), that is, four sets of subjects' weights and four sets of response-option weights. However, it is well known that subject's scores and response-option's weights do not span the same space. This discrepancy can be eliminated by doubling the space dimensionality (Nishisato 1980). In other words, by expanding space dimensionality from four to eight, we can introduce the coordinate system which allows us to plot both column variables and row variables in a common space. This corresponds to the analysis of the super-distance matrix (Nishisato and Clavel 2010) as shown in Table 2. This space that accommodates both within-set distances and between-set distances is called *doubled hyper-space* in the current paper.

Table 2 Super distance matrix

	Options of Item 1	Options of Item 2
Options of Item 1	Within-set distance	Between-set distance
Options of Item 2	Between-set distance	Within-set distance

Table 3 Contingency table of Items 1 and 2

		Item 2 Option 1	Item 2 Option 2	Item 2 Option3
Item 1	Option 1	0	1	3
Item 1	Option 2	1	1	0
Item 1	Option 3	1	2	1

4 Vindication of the Carroll-Green-Schaffer Scaling

Let us consider the two questions in our example. Their response-pattern representation is a 10×6 matrix, corresponding to Items 1 and 2 of the nominal-scale format in Table 1. This data set can also be expressed as the options-by-options contingency table (Table 3). In the latter form, it is known that the options of Question 1 and those of Question 2 (i.e. between-set distances) do not span the same space. The Carroll-Green-Schaffer scaling (Carroll et al 1986) transforms the contingency table format to the 10×6 response-pattern format so that the six options are now all column variables (i.e., within-set variables), thus span the same space. Greenacre (1989) criticized the Carroll-Green-Schaffer scaling as incorrect. Notice, however, that Nishisato (1980) presented a thorough comparisons of scaling these two formats, showing (1) the two sets of option weights for the contingency table are identical to the first two sets of weights from the response-pattern matrix, and (2) the response-pattern format yields two additional components (see Nishisato 1980, for a numerical example). This point was written by Nishisato, after seeing Greenacre's criticism, but the editor of that journal rejected Nishisato's paper without review, stating that the matter had already been resolved in Greenacre's favor. Nishisato's point then was that the six response options could be accommodated in the same space, not in 2-dimensional space, but in 4-dimensional space, the idea of doubled hyper-space. In other words, one can place two sets of options in columns, subject the data to quantification, and plot their positions in the same space, provided that we double the dimensionality of space. In conclusion, the Carroll-Green-Schaffer scaling is correct, so long as we double the dimensionality of space. Furthermore, if we wish to accommodate not only all options of two items, but also all subjects in common space, we need 8-dimensional space (i.e., additional discrepancies between the space for subjects and the space for items require the added dimensionality).

5   Cluster Analysis of Variables Through Filters

Multidimensional analysis of categorical data in doubled hyper-space was the view advanced by total information analysis or comprehensive dual scaling (Nishisato and Clavel 2010). Their analysis starts with extracting all components, and constructs the super-distance matrix, consisting of within-set and between-set distances, using all components. The idea behind their insistence of using all components lies in the hope to view the undistorted configuration of data. Due to the doubled dimensionality of space, they proposed to apply the k-means clustering to the complete information contained in this super-distance matrix. Nishisato (2012b) suggested cluster analysis of only the between-set distance matrix since our primary interest lies in the relation between row variables and column variables, and proposed a simple minimum-distance clustering procedure, in which only the minimum distance in each row and the minimum distance in each column are retained for identifying associations.

The current paper proposes yet another procedure of clustering, called cluster analysis through filters. From the complete between-set distance matrix, the p-percentile distance is calculated, which is then used to filter the distances. We discard all distances greater than the p-percentile point, and from the remaining distances we identify clusters of variables. The percentile point can be changed to identify tighter or looser clusters. This procedure is purely descriptive and does not involve the problem of identifying the number of clusters or mutually exclusive clusters. As a few preliminary investigations have demonstrated, the best value of p depends on data sets and on the nature of clusters, that is, mutually exclusive versus overlapping. Thus, the problem of choosing the best value of p will be left for future research.

6   Example: Heuer's Suicide Data

Heuer's suicide data (Heuer 1979) are often used in psychometrics (e.g. Van der Heijden and De Leeuw 1985; Nishisato 1994). The data are frequencies of 9 types of suicides collected from 17 age groups of each of the 2 gender groups (Table 4). It is known that the data contain 2 major components which account for 90 % of the total information. Therefore, the results are typically graphed in two dimensions (Fig. 1), using symmetric scaling (i.e., row variables and column variables are graphed in two dimensions, as if they span the same space) and conclusions are drawn directly from the graph. To be exact, however, the relations among those variables in this 2-dimensional graph requires a 4-dimensional space to account for 90 % of information, and our 2-dimensional graph is a distortion of the correct configuration. More specifically, the discrepancies in angles between the axis for row variables and

Table 4 Ages, genders and forms of suicides

	sm	gsH	gsO	hng	drwn	gun	knf	jmp	oth
Male 10–	4	0	0	247	1	17	1	6	9
16–	348	7	67	578	22	179	11	74	175
21–	808	32	229	699	44	316	35	109	289
26–	789	26	243	648	52	268	38	109	226
31–	916	17	257	825	74	291	52	123	281
36–	1,118	27	313	1,278	87	293	49	134	268
41–	926	13	250	1,273	89	299	53	78	198
46–	855	9	203	1,381	71	347	68	103	190
51–	684	14	136	1,282	87	229	62	63	146
56–	502	6	77	972	49	151	46	66	77
61–	516	5	74	1,249	83	162	52	92	122
66–	513	8	31	1,360	75	164	56	115	95
71–	425	5	21	1,268	90	121	44	119	82
76–	266	4	9	866	63	78	30	79	34
81–	159	2	2	479	39	18	18	46	19
86–	70	1	0	259	16	10	9	18	10
90+	18	0	1	76	4	2	4	6	2
Female 10–	28	0	3	20	0	1	0	10	6
16–	353	2	11	81	6	15	2	43	47
21–	540	4	20	111	24	9	9	78	67
26–	454	6	27	125	33	26	7	86	75
31–	530	2	29	178	42	14	20	92	78
36–	688	5	44	272	64	24	14	98	110
41–	566	4	24	343	76	18	22	103	86
46–	716	6	24	447	94	13	21	95	88
51–	942	7	26	691	184	21	37	129	131
56–	723	3	14	527	163	14	30	92	92
61–	820	8	8	702	245	11	35	140	114
66–	740	8	4	785	271	4	38	156	90
71–	624	6	4	610	244	1	27	129	46
76–	495	8	1	420	161	2	29	129	35
81–	292	3	2	223	78	0	10	84	23
86–	113	4	0	83	14	0	6	34	2
90+	24	1	0	19	4	0	2	7	0

Notes: sm = splidmatters, gsH = gas at home, gsO = gas others, hng = hanging, drwn = drowning, gun = guns and explosives, knf = knifing, jmp = jumping, oth = other means

the axis for column variables (Nishisato and Clavel 2003) are 71.8° for component 1 and 74.5° for component 2, both of which are definitely not small enough to ignore. Since a 4-dimensional graph is beyond our ability to present here, we must resort to just pointing out the problem inherent in our traditional 2-dimensional so-called symmetric graph. This graph is useful only if these discrepancy angles are minimal, which is not the case with the current example.

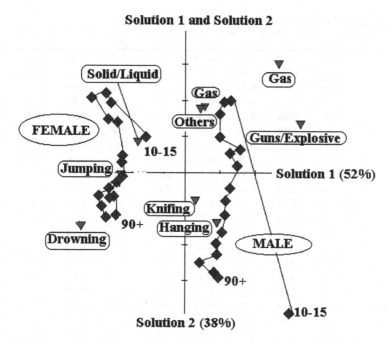

Fig. 1 Two-dimensional symmetric scaling plot of Heuer's data

In our approach, we first extracted all eight components, in terms of which the between-set distances were calculated (Table 5, reproduced here from Nishisato 2012b). Cluster analysis of the between-set distance matrix is then subjected to 2 filters, the 50 percentile and the 25 percentile filters (Table 6). These filters were chosen just for the purpose of demonstration, not based on any rationale. We can immediately see that

- There are predominant suicide types: Use of solid and liquid matters (drugs, poisons), hanging, knives, jumping and other means.
- There are gender and age related types: young males and gas, old females and drowning, males and guns/explosives.

This is a summary of the actual distribution of information. If we see any discrepancies between the symmetric graph and the filtered distances, the latter is always closer to the information in the data. Wouldn't this simple summary be more useful and informative than the traditional grossly distorted analysis through dimension reduction? The traditional 2-dimensional graph looks beautiful, but it has a serious trap that it is an overlay of two separate 2-dimensional configurations onto the same space, making a direct interpretation totally illogical.

Table 5 Age and suicide between-set distance table

	sm	gsH	gsO	hng	drwn	gun	knf	jmp	oth
Male 10–	1.21	1.29	1.39	0.90	1.35	1.17	0.99	1.21	1.20
16–	0.60	0.68	0.81	0.51	1.00	0.66	0.51	0.69	0.50
21–	0.60	0.63	0.74	0.64	1.04	0.68	0.60	0.72	0.48
26–	0.57	0.62	0.73	0.62	1.01	0.68	0.57	0.70	0.47
31–	0.51	0.60	0.73	0.56	0.97	0.66	0.51	0.65	0.42
36–	0.47	0.59	0.74	0.46	0.92	0.65	0.42	0.60	0.41
41–	0.50	0.62	0.76	0.42	0.92	0.64	0.39	0.62	0.45
46–	0.53	0.65	0.78	0.39	0.93	0.63	0.38	0.63	0.48
51–	0.53	0.67	0.83	0.33	0.89	0.67	0.34	0.61	0.51
56–	0.54	0.70	0.88	0.32	0.88	0.70	0.34	0.60	0.55
61–	0.57	0.74	0.92	0.32	0.87	0.73	0.36	0.62	0.59
66–	0.63	0.80	0.99	0.37	0.89	0.79	0.41	0.65	0.66
71–	0.65	0.83	1.03	0.39	0.88	0.84	0.44	0.66	0.70
76–	0.70	0.88	1.07	0.44	0.90	0.88	0.49	0.70	0.75
81–	0.70	0.91	1.12	0.47	0.87	0.94	0.51	0.69	0.79
86–	0.78	0.96	1.15	0.52	0.94	0.96	0.58	0.77	0.85
90+	0.83	1.00	1.18	0.57	0.98	0.99	0.62	0.82	0.89
Female 10–	0.57	0.81	1.04	0.73	0.94	0.98	0.67	0.61	0.66
16–	0.6	0.90	1.12	0.91	1.00	1.12	0.83	0.71	0.75
21–	0.62	0.91	1.15	0.92	0.96	1.16	0.83	0.70	0.77
26–	0.52	0.80	1.05	0.81	0.90	1.04	0.72	0.60	0.65
31–	0.48	0.80	1.06	0.77	0.85	1.04	0.67	0.56	0.64
36–	0.42	0.75	1.01	0.71	0.82	0.99	0.62	0.52	0.58
41–	0.35	0.72	1.01	0.59	0.73	0.95	0.50	0.42	0.56
46–	0.36	0.74	1.02	0.59	0.72	0.96	0.50	0.43	0.58
51–	0.37	0.75	1.04	0.55	0.68	0.96	0.47	0.42	0.59
56–	0.41	0.78	1.07	0.58	0.67	0.99	0.50	0.45	0.63
61–	0.46	0.81	.11	0.59	0.66	1.03	0.52	0.47	0.67
66–	0.52	0.85	1.15	0.60	0.66	1.05	0.54	0.51	0.73
71–	0.58	0.90	1.19	0.66	0.68	1.11	0.61	0.56	0.79
76–	0.54	0.87	1.18	0.67	0.68	1.10	0.60	0.52	0.76
81–	0.52	0.86	1.17	0.68	0.70	1.10	0.61	0.51	0.75
86–	0.57	0.87	1.17	0.71	0.80	1.10	0.65	0.56	0.77
90+	0.59	0.88	1.19	0.71	0.78	1.11	0.64	0.57	0.79

In practice, one should try other filters as well, and choose the one best suited for the interpretation of data. If all variables are to be included in clustering, the minimum distances of individual rows and columns (Nishisato 2012b) will guarantee it. Another way of making sure that all variables are equally involved in clustering is to consider *conditional filtering* such that the minimum three distances, for example, are kept in each row and each column.

Table 6 Sifting with filters 50 and 25 percentiles

Gender-Age	sm	gsH	gsO	hng	drwn	gun	knf	jmp	oth
M 10–15				90*					
M 16–	(60)	68		(51)		66	(51)	69	(50)*
M 21–	(60)	63		64		68	(60)		(48)*
M 26–	(57)	62	73*	62		68	(57)	70	(47)*
M 31–	(51)	(60)	73*	(56)		66	(51)	65	(42)*
M 36–	(47)	(59)*		(46)		65	(42)	(60)	(41)*
M 41–	(50)	62		(42)		64	(39)*	62	(45)
M 46–	(52)	65		(39)		63*	(38)*	63	(48)
M 51–	(53)	67		(33)*		67	(34)	61	(51)
M 56–	(54)	70		(32)*		70	(34)	(60)	(55)
M 61–	(57)			(32)*			(36)	62	(59)
M 66–	63			(37)*			(41)	65	66
M 71–	65			(39)*			(44)	66	70
M 76–	70			(44)*			(44)	66	70
M 81–	70			(47)*			(51)	(60)	
M 86–				(52)*			(58)		
M 90–				(57)*			62		
F 10–	(57)*						67	61	66
F 16–	62*								
F 21–	62*							70	
F 26–	(52)*							(60)	65
F 31–	(48)*						67	(56)	64
F 36–	(42)*						62	(52)	(58)
F 41–	(35)*			(59)			(60)	(42)	(56)
F 46–	(36)*			(50)			(50)	(43)	(58)
F 51–	(37)*			(55)	68		(47)	(42)	(59)
F 56–	(41)*			(58)	67		(50)	(45)	63
F 61–	(46)*			(59)	66		(52)	(47)	67
F 66–	(52)			(60)	66		(54)	(51)*	
F 71–	(58)			66	68		61	(56)*	
F 76–	(54)			68	68		69	(52)*	
F 81–	(52)			68	70		61	(51)*	
F 86–	(57)						65	(56)*	
F 90–	(59)						64	(57)*	

Note 1: M = male; F = female
Note 2: Regular numbers indicate results by the 50 percentile filter, (**bold-faced numbers in parentheses**) by the 25 percentile filter, and the asterisks* by minimal values in each row and column

7   Concluding Remarks

The idea of doubled hyper-space for data analysis places a heavy onus on the researchers if they are to follow the traditional multidimensional graphic approach. The present study offers a simple alternative procedure of cluster analysis through filters, which makes it possible to deal with doubled hyper-space without distorting the configuration of multidimensional data. It is a descriptive way of deciphering a complex cloud of data points, and offers some promise even for analysis of exceptionally high-dimensional data. Cluster analysis is a dimension-free process, and is seemingly suited to deal with doubled hyper-space.

Acknowledgements Thanks are due to José Garcia Clavel for the calculation of between-set distances of Heuer's data.

References

Carroll JD, Green PE, Schaffer CM (1986) Interpoint distance comparisons in correspondence analysis. J Mark Res 23(3):271–280

Greenacre MJ (1989) The carroll-green-schaffer scaling in correspondence analysis: a theoretical and empirical appraisal. J Mark Res 26(3):358–365

Heuer G (1979) Selbstmord bei Kindern und Jugendlichen: ein Beitrag zur Suizidprophylaxe aus pädagogischer Sicht. Klett-Cotta, Stuttgart

Likert R (1932) A technique for the measurement of attitudes. Arch Psychol 22(140):44–53

Nishisato S (1980) Analysis of categorical data: dual scaling and its applications. University of Toronto Press, Toronto

Nishisato S (1984) Forced classification: a simple application of a quantification method. Psychometrika 49:25–36

Nishisato S (1994) Elements of dual scaling: an introduction to practical data analysis. Lawrence Erlbaum Associates, Hillsdale

Nishisato S (1999) Data types and information: beyond the current practice of data analysis. In: Decker R, Gaul W (eds) Classification and information processing at the turn of the Millennium. Springer, Berlin/Heidelberg, pp 40–51

Nishisato S (2006) Correlational structure of multiple-choice data as viewed from dual scaling. In: Greenacre MJ, Blasius I (eds) Multiple correspondence analysis and related methods. Chapman and Hall/CRC, Boca Raton, chap 6, pp 161–178

Nishisato S (2007) Multidimensional nonlinear descriptive analysis. Chapman and Hall/CRC, Boca Raton

Nishisato S (2012a) Optimal quantities for analysis through regression of measurement on data. Bull Data Anal Jpn Classif Soc 1:1–10

Nishisato S (2012b) Reminiscence and a step forward. In: Gaul W, Geyer-Schultz A, Schmidt-Thieme L, Kunze J (eds) Classification, data analysis, and knowledge organization. Springer, Heidelberg, pp 109–119

Nishisato S, Baba Y (1999) On contingency, projection and forced classification of dual scaling. Behaviormetrika 26:207–219

Nishisato S, Clavel JG (2003) A note on between-set distances in dual scaling and correspondence analysis. Behaviormetrika 30(1):87–98

Nishisato S, Clavel JG (2010) Total information analysis: comprehensive dual scaling. Behaviormetrika 57:15–32

Van der Heijden PGM, De Leeuw J (1985) Correspondence analysis used complementary to loglinear analysis. Psychometrika 50(4):429–447

Three-Mode Hierarchical Subspace Clustering with Noise Variables and Occasions

Kensuke Tanioka and Hiroshi Yadohisa

Abstract Three-mode data are observed in various domains such as panel research in psychology studies. To conceive clustering structures from three-mode data as an initial analysis, a clustering algorithm is applied to the data. However, traditional clustering algorithms cannot factor in the effects of occasions. In addition, it is difficult to understand these typically high-dimensional data. Although Vichi et al. (J Classif 24(1):71–98, 2007) proposed three-way clustering, their algorithms are based on complicated assumptions.

We propose three-mode subspace clustering based on entropy weights. The proposed algorithm excludes complicated assumptions and provides results that can be easily interpreted.

1 Introduction

Three-mode data are described as $\bar{X} \in \mathbb{R}^{|I| \times |J| \times |K|}$, where I, J and K represent a set of objects, variables, and occasions respectively, and $| \cdot |$ represents the cardinality of sets. Three-mode data are observed in various domains such as marketing science, Web mining, and psychology. For example, in panel research, I, J and K are described as a set of participants, a set of questions, and a set of years, respectively. Also, there are many situations in which the clustering algorithm can be applied to three-mode data to reveal the clustering structures of

K. Tanioka
Graduate school of Culture and Information Science, Doshisha University, Kyoto 610-0313, Japan
e-mail: eim1001@mail4.doshisha.ac.jp

H. Yadohisa (✉)
Department of Culture and Information Science, Doshisha University, Kyoto 610-0313, Japan
e-mail: hyadohis@mail.doshisha.ac.jp

W. Gaul et al. (eds.), *German-Japanese Interchange of Data Analysis Results*,
Studies in Classification, Data Analysis, and Knowledge Organization,
DOI 10.1007/978-3-319-01264-3__8,
© Springer International Publishing Switzerland 2014

objects. In such cases, typically, three-mode data are converted into two-mode data $X_{I,JK} = (X_{..1}, X_{..2}, \cdots, X_{..K})$ where $X_{..k} \in \mathbb{R}^{|I| \times |J|}$ is described as two-mode data under the occasion k ($\in K$) and is applied to the clustering algorithm for two-mode data. However, such a strategy leads to three problems. First, $X_{I,JK}$ tends to become high dimensional data since $X_{I,JK}$ has a number of $|J| \times |K|$ variables. Therefore, this approach leads to the curse of dimensionality (Bishop 2006). Second, it is difficult to interpret the clustering results from high-dimensional data. Finally, we note that the strategy does not factor in occasions.

To overcome these problems, subspace clustering algorithms may be applied to the data $X_{I,JK}$. There are two kinds of subspace clustering approaches. In the first approach, subspace variables are described as a few new variables derived from a linear combination of the original variable spaces with strong assumptions. For three-mode data, Vichi et al. (2007) proposed Tucker3 Clustering (T3Clus) and Three-way Factorial k-means (3Fk-means) based on this approach. These algorithms consider the factors of variables and occasions, respectively provide the clustering results, and simultaneously visualize features for the clusters. However, by this approach, it is difficult to interpret the results and eliminate the effects of noise variables (Parsons et al. 2004), where noise variables are defined as variables that do not include clustering structures and mask the true clustering structures. In addition to that, these algorithms assume that the clusters are hyperspherical and each cluster has the same number of objects since these algorithms are based on the least-squares criterion. In the other approach, a subspace is described as a subset of V, K for each cluster so that these variables and occasions have clustering structures through the weights of variables and occasions for each cluster are estimated (Parsons et al. 2004). The approach is simple, and hence it is easy to interpret the clustering results. However, subspace clustering based on the three-mode data approach has not been proposed so far. For two-mode data, clustering objects on subsets of attributes (COSA), which is a subspace clustering algorithm based on this approach, has been proposed by Friedman and Meulman (2004). In COSA, it is easy to interpret the model based on entropy for which the subspace is constructed.

Then, in this study, we propose a new hierarchical subspace clustering algorithm for three-mode data based on COSA. The proposed algorithm consists of two steps. In the first step, weights for variables and occasions on each cluster are estimated such that effects of noise variables are reduced. In the second step, two clusters are combined based on the minimum distance rule. These steps are iterated till the number of clusters becomes one. The proposed algorithm has several advantages for classifying objects from three-mode data. First, the proposed algorithm can detect clusters on subsets of variables and occasions. In short, the proposed algorithm eliminates the effects of masking variables. Then, researchers tend to easily conceive the features of each cluster. Finally, the proposed algorithm does not assume that each cluster has the same number of objects and does not determine the number of clusters beforehand since it is a hierarchical clustering algorithm.

2 Objective Function and Update Formula

In this section, we introduce a new objective function for the proposed hierarchical clustering algorithm and define an update formula for weight functions of variables and occasions.

2.1 Objective Function

Here we define the proposed objective function for three-mode subspace clustering based on COSA.

Definition 1 (Objective function). When $X_{I,JK}$, parameters $\lambda(\geq 0)$ and $\eta(\geq 0)$ (these parameters control the scale of weights for variables and occasions, respectively) are given as input parameters, the objective function Q and the optimization problem are defined as follows:

$$Q(\{w_g\}_1^{|G|}, \{z_g\}_1^{|G|}) = \sum_{g \in G} \left\{ (u_g^T u_g (u_g^T u_g - 1))^{-1} (w_g \otimes u_g)^T \bar{D} (z_g \otimes u_g) + \right.$$

$$\left. |K|\lambda w_g^T \log(w_g) + |J|\eta z_g^T \log(z_g) + |K|\lambda \log(|J|) + |J|\eta \log(|K|) \right\}$$

$$= \sum_{g \in G} \frac{1}{|C_g|(|C_g| - 1)} \sum_{s \in C_g} \sum_{t \in C_g} D_{st}^{(\lambda,\eta)}[w_g, z_g] \to Min$$

$$D_{st}^{(\lambda,\eta)}[w_g, z_g] = \sum_{j \in J} \sum_{k \in K} \{w_{kg} z_{kg} d_{stjk} + \lambda w_{jg} \log(w_{jg}) + \eta z_{kg} \log(z_{kg})\}$$

$$+ |K|\lambda \log(|J|) + |J|\eta \log(|K|)$$

subject to:

$$\sum_{j \in J} w_{jg} = 1, \quad w_{jg} \geq 0, \quad \sum_{k \in K} z_{kg} = 1, \quad z_{kg} \geq 0, \tag{1}$$

where, $\otimes$ indicates the Kronecker product, G and $C_g, (g \in G)$ represent set of indices for each cluster and clusters, $w_g = (w_{1g}, w_{2g}, \cdots, w_{|J|g})^T$ and $z_g = (z_{1g}, z_{2g}, \cdots, z_{|K|g})^T$ respectively, indicate the weights of variables j and occasions k on cluster C_g, respectively, $u_g = (u_{1g}, u_{2g}, \cdots, u_{|I|g})^T$ show the indicator vectors, $u_{ig} = 1$ if object i belongs to C_g else $u_{ig} = 0$, $\log(w_g) = (\log(w_{1g}), \log(w_{2g}), \cdots, \log(w_{|J|g}))^T$, $\log(z_g) = (\log(z_{1g}), \log(z_{2g}), \cdots, \log(z_{|K|g}))^T$, and

$$\bar{D} = \begin{bmatrix} D_{11} & D_{12} & \cdots & D_{1|K|} \\ D_{21} & D_{22} & \cdots & D_{2|K|} \\ \vdots & \vdots & \ddots & \vdots \\ D_{|J|1} & D_{|J|2} & \cdots & D_{|J||K|} \end{bmatrix},$$

$D_{jk} = (d_{stjk})$ represents the dissimilarity matrix between objects $(s, t \in I)$ for variable j and occasion k.

$w_{jg} \log(w_{jg})$ and $z_{kg} \log(z_{kg})$ are penalty terms for the weights of variables and of occasions respectively, and $|K|\lambda \log(|J|)$ and $|J|\lambda \log(|K|)$ are added to satisfy $D^{(\lambda,\eta)}[w_g, z_g] \geq 0$ for Q.

The objective function provides large weight to the variable j and occasion k on C_g whose average distance of $s, t \in C_g$ is small; in short, we consider variable j and occasions k on C_g whose average distance of $s, t \in C_g$ is large as masking variables and the weight become small.

2.2 Update Formula for Weights of Variables and Occasions

In this subsection, we show the two propositions for updating formulas of w_g and z_g and $D^{(\lambda,\eta)}[w_g, z_g]$ minimizing the objective function Q. The proposed algorithm adopts an alternative least squares fashion approach. Precisely, updating w_g and z_g for $g \in G$ is iterated and the estimated weights for both w_g and z_g are guaranteed for monotone non-increasing objective functions, because each regularized term of the objective function Q is a convex function.

Proposition 1 (Update formula for weighs of variables j on C_g, w_g). *When Q and z_g are given, w_g, $(g \in G)$, minimizing Q subject to (1) is described as follows:*

$$w_g = \frac{\exp\left\{\left(-|K|\lambda u_g^T u_g (u_g^T u_g - 1)\right)^{-1} (E_J \otimes u_g)^T \bar{D}(z_g \otimes u_g)\right\}}{1_J^T \exp\left\{\left(-|K|\lambda u_g^T u_g (u_g^T u_g - 1)\right)^{-1} (E_J \otimes u_g)^T \bar{D}(z_g \otimes u_g)\right\}},$$

$$w_{jg} = \frac{\exp\left\{(-|K|\lambda)^{-1} S_{jg}^{(K)}\right\}}{\sum_{j^* \in J} \exp\left\{(-|K|\lambda)^{-1} S_{j^* g}^{(K)}\right\}}, \quad S_{jg}^{(K)} = \frac{1}{|C_g|(|C_g| - 1)} \sum_{s \in C_g} \sum_{t \in C_g} \sum_{k \in K} z_{kg} d_{stjk},$$

where $E_J = (e_{pq})_{|J| \times |J|}$ is a unit matrix, $exp\{x\} = (\exp\{x_1\}, \exp\{x_2\}, \cdots, \exp\{x_n\})$, $x \in \mathbb{R}^n$, and 1_J represents a vector that consists of 1 with length of $|J|$.

Next, we show the update formula for z_g

Proposition 2 (Update formula for weighs of occasions k on C_g, z_g). *When Q and w_g are given, z_g, $(g \in G)$, minimizing Q subject to (1) is described as follows:*

$$z_g = \frac{\exp\left\{\left(-|J|\eta u_g^T u_g (u_g^T u_g - 1)\right)^{-1} (E_K \otimes u_g)^T \bar{D}^T (w_g \otimes u_g)\right\}}{1_K^T \exp\left\{\left(-|J|\eta u_g^T u_g (u_g^T u_g - 1)\right)^{-1} (E_K \otimes u_g)^T \bar{D}^T (w_g \otimes u_g)\right\}},$$

$$z_{kg} = \frac{\exp\left\{(-|J|\eta)^{-1} S_{kg}^{(J)}\right\}}{\sum_{k^* \in K} \exp\left\{(-|J|\eta)^{-1} S_{k^* g}^{(J)}\right\}}, \quad S_{kg}^{(J)} = \frac{1}{|C_g|(|C_g| - 1)} \sum_{s \in C_g} \sum_{t \in C_g} \sum_{j \in J} w_{jg} d_{stjk},$$

where $E_K = (e_{pq})_{|K| \times |K|}$ is a unit matrix and 1_L represents a vector that consists of 1 with a length of $|K|$.

2.3 Dissimilarity Between Clusters

To determine a combined criterion between clusters based on Q, we introduce dissimilarities based on COSA. The similarity between clusters come to be asymmetry because each cluster has different weights for variables and occasions. Then, the asymmetric similarity is converted into a symmetric similarity.

Definition 2 (Dissimilarities for three-mode data between clusters). When w_g and z_g are calculated by an update formula, dissimilarities used for the proposed algorithm are defined as follows:

$$\mathscr{D}(C_g, C_h | w_g, w_h, z_g, z_h)^{(1)} = u_g^T \max(\mathscr{D}(C_g, C_h | w_g, z_g), \mathscr{D}(C_g, C_h | w_h, z_h)) u_h, \tag{2}$$

$$\mathscr{D}(C_g, C_h | w_g, w_h, z_g, z_h)^{(2)} = (u_g^T u_g u_h^T u_h)^{-1} \left\{ (w_{gh}^\# \otimes u_g)^T \bar{D}(z_{gh}^\# \otimes u_h) \right.$$
$$+ |K| w_{gh}^{\#T} \log(w_{gh}^\#) + |J| z_{gh}^{\#T} \log(z_{gh}^\#)$$
$$\left. + |K| \lambda \log(|J|) + |J| \eta \log(|K|) \right\}, \tag{3}$$

where

$$\mathscr{D}(C_g, C_h | w_g, z_g) = (u_g^T u_g u_h^T u_h)^{-1} \left\{ (w_g \otimes diag\{u_g\})^T \bar{D}(z_g \otimes diag\{u_h\}) \right.$$
$$+ |K| w_g^T \log(w_g) u_g u_h^T + |J| z_g^T \log(z_g) u_g u_h^T + |K| \lambda \log(|J|) u_g u_h^T$$
$$\left. + |J| \eta \log(|K|) u_g u_h^T \right\},$$

$$\max(A, B) = (\max(a_{st}, b_{st}))_{m \times n}, \quad A, B \in \mathbb{R}^{m \times n},$$

$$w_{g,h}^\# = (\max(w_{1g}, w_{1h}), \max(w_{2g}, w_{2h}), \cdots, \max(w_{|J|g}, w_{|J|h})),$$

$$z_{g,h}^\# = (\max(z_{1g}, z_{1h}), \max(z_{2g}, z_{2h}), \cdots, \max(z_{|K|g}, z_{|K|h})).$$

$\mathscr{D}^{(1)}$ and $\mathscr{D}^{(2)}$ are used to combine two clusters based on the minimum distance rule, and the combination leads to Q minimized for the greedy minimization strategy.

3 Algorithm

In this section, we discuss the proposed algorithm. For agglomerative hierarchical clustering, initial clusters are described as singleton clusters. Then, w_g and z_g of the initial clusters cannot be calculated since these updated formulae need more than two objects in each cluster.

To overcome the problem for the initial weights, two approaches are determined. For the first approach, when the variation of subclusters (clusters with a small number of objects), is less than the thresholds obtained by researchers, w_g and z_g are calculated on the basis of subclusters and are generated from $X_{I,JK}$ by using BIRCH (Zhang et al. 1996) as the initial clusters. Although this approach can be applied to large data, the approach cannot be applied if the number of objects is small. For the other approach, the k-nearest neighbor KNN is used for $X_{I,JK}$ with the number of objects being small. The approach is defined by Friedman and Meulman (2004).

Definition 3 (Determining the initial KNN indicator matrix). When $X_{I,JK}$ and a parameter of the KNN, $p \in \{1, 2, \cdots, |I|\}$, are given, the KNN indicator vector $v_i = (v_{i1}, v_{i2}, \cdots, v_{i|I|})$ is defined as follows:

$$v_{is} = \begin{cases} 1 & (s \in KNN(i)) \\ 0 & (otherwise), \end{cases}$$

$KNN(i)$ indicates $\{s| \ D_{is}^* \leq D_{i(p)}\}$, $D_{is}^* = (1_J \otimes e_i)^T \bar{D}(1_K \otimes e_s)$, $E_I = (e_1, e_2, \cdots, e_{|I|})$ is a unit matrix and $e_i \in \mathbb{R}^{|I|} (i \in I)$ represents a column vector of E_I and $D_{i(p)}$ represents the order statistics of $\{D_{is}\}_{s=1}^{|I|}$.

Then, based on Definition 3, the algorithm for calculating initial weights w_i, z_i is presented as Algorithm 1, and the proposed subspace clustering for three-mode data is shown as Algorithm 2. Algorithm 1 consist of two steps. First, a KNN indicator vector for each object is calculated from $X_{I,JK}$ and p. Second, Calculating weights for variable j and occasion k on $V_i = \{i| \ v_{is} = 1, i \in I, s \in KNN(i)\}$ based on like the alternative least squares fashion till a stop rule is satisfied. For Algorithm 2, it also consists of two steps. First, weights for variable j and occasion k on clusters $C_g (g \in G)$ are calculated. Second, two clusters are combined based on minimum distance rule. Then, these steps are iterated till the stop rule is satisfied.

We define some notations to describe Algorithms 1 and 2 (see Fig. 1) such that $^{(t)}Q = {}^{(t)}Q(\{w_g\}_1^{|G|}, \{z_g\}_1^{|G|})$, $\mathscr{D}(z_{g*}^{(q)}, z_{h*}^{(q)}|C_g^*, C_h^*, w_g, w_h)$ represents z corresponding to $\mathscr{D}^{(\ell)}(C_g^*, C_g^*|w_g^*, w_h^*, z_g^*, z_h^*)$, $(\ell \in \{1, 2\})$ and $Q_g = (u_g^T u_g (u_g^T u_g - 1))^{-1}(w_g \otimes u_g)^T \bar{D}(z_g \otimes u_g) + |K|\lambda w_g^T \log(w_g) + |J|\eta z_g^T \log(z_g) + |K|\lambda \log(|J|) + |J|\eta \log(|K|)$.

Algorithm 1: Calculating initial weights
1: Set initial weights $w_i \leftarrow 1/|J|$ and $z_i \leftarrow 1/|K|$
2: Set initial parameters λ, η
3: Set $\varepsilon > 0$ to a small positive number
4: **for** $i = 1$ to $|I|$ **do**
5:    Compute $KNN(i)$ and ν_i
6: **end for**
7: $t \leftarrow 1$
8: **While** $|^{(t)}Q - ^{(t-1)}Q| > \varepsilon$ **do**
9:    **for** $i = 1$ to $|I|$ **do**
10:     **for** $j = 1$ to $|J|$ **do**
11:      Update $^{(t+1)}w_i$
12:     **end for**
13:     **for** $k = 1$ to $|K|$ **do**
14:      Update $^{(t+1)}z_i$
15:     **end for**
16:    **end for**
17: $t \leftarrow t + 1$
18: **end while**

Algorithm 2 : Proposed subspace clustering for three-mode data
1: Set initial weights $w_g, z_g, (g \in G)$, $t \leftarrow 1$ and $q \leftarrow 1$
2: Set initial indicator vectors $u_g, (g \in G)$
3: Set initial parameters λ, η
4: Set $\varepsilon > 0$ to a small positive number
5: **While** $u_{g*}^T u_{g*} = |I|$ **do**
6:    Compute $\{g^*, h^*\} = \mathrm{argmin}_{g,h} \mathscr{D}(C_g, C_h | w_g, w_h, z_g, z_h)$
7:    **for** $g = 1$ to $|G|$ **do**
8:     $u_g^{(q+1)} \leftarrow u_g^{(q)}$; $w_g^{(q+1)} \leftarrow w_g^{(q)}$; $z_g^{(q+1)} \leftarrow z_g^{(q)}$
9:    **end for**
10:    $u_{g*}^{(q+1)} \leftarrow u_{g*}^{(q)} + u_{h*}^{(q)}$
11:    $z_{g*}^{(q+1)} \leftarrow \mathscr{D}(z_{g*}^{(q)}, z_{h*}^{(q)} | C_g^*, C_h^*, w_g^{(q)}, w_h^{(q)})$
12:    $q \leftarrow q + 1$; $G \leftarrow \{g | h^* \neq g, g \in G\}$
13:    set $z_{g*} \leftarrow 1_K/|K|$
14:    **While** $|^{(t)}Q_{g*} - ^{(t-1)}Q_{g*}| > \varepsilon$ **do**
15:     **for** $j = 1$ to $|J|$ **do**
16:      Update $^{(t+1)}w_{g*}^{(q)}$
17:     **end for**
18:     **for** $k = 1$ to $|K|$ **do**
19:      Update $^{(t+1)}z_{g*}^{(q)}$
20:     **end for**
21:     $t \leftarrow t + 1$
22:    **end while**
23: **end while**

Fig. 1 Algorithms 1 and 2

4 Numerical Example

In this section, we show the numerical example using "background to the Electronics Industries Data" (Ambra 1985). The data consist of 24 countries, 6 industries, and 7 years as I, J, and K, respectively. The variables consist of information science (INFO), radio and television equipment (RADI), telecommunications products (TELE), electromedical equipment (ELET), scientific equipment and components and parts (COMP). Similarly, the occasions are observed each year from 1978 to 1985 but excluding 1981. Each cell shows a specialization index and these values are determined as "the proportion of the monetary value of an electronic industry compared to the total export value of manufactured goods of a country compared to the similar proportion for the world as whole" (Ambra 1985).

We apply the proposed clustering algorithm to the data with $\lambda = 1/35$, $\eta = 1/60$, and $p = 3$. Then, we interpret the clustering results, for when there are three clusters, it is easy to interpret the results. We show the clustering results and weights of variables and occasions in Table 1 and Fig. 2.

From Fig. 2, each cluster is embedded in a differential subspace of variables and occasions from the other clusters. Cluster 1 tends to have similar values for variables of RADI and occasions of the old years under the period. Objects in cluster 2 are mainly connected in INFO for the variables, and the middle of the period for the occasions. For cluster 3, there are INFO, RADI, and COMP for variables and the end of the period for the occasions.

Table 1 Clustering results for "background to the Electronics Industries Data"

I	CA	US	AS	DA	FR	RF	IR	IT	PB	RU	NO	SP	SV	YU	JP	BL	AU	FI	PO	NZ	GR	CH	TU
G	1	1	1	1	1	1	1	1	1	1	1	1	1	1	1	1	2	2	2	2	2	3	3	3	3

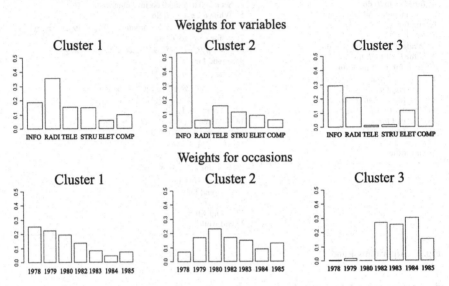

Fig. 2 Weights of variables and occasions for each cluster

5   Conclusion

In this paper, we proposed hierarchical subspace clustering for three-mode data for considering the effects of variables and occasions. The proposed method provides variables to reduce the effects of masking variables and features of clusters through the weight for variables and occasions. In addition to that, it is easy to understand the clustering results since the proposed algorithm does not include complicated assumptions. For future study, we aim to optimize the objective function with λ and η as COSA.

References

Ambra L (1985) Alcune estensione dell'analysi in componenti principali per lo studio dei sistemi evolutivi. uno studio sul commercio internazionale dell'elettronica. Rivista Economica del Mezzogiorno 2:233–260

Bishop CM (2006) Pattern recognition and machine learning. Springer, Berlin

Friedman JH, Meulman JJ (2004) Clustering objects on subsets of attributes. J R Stat Soc: Ser B (Stat Methodol) 66(4):815–849

Parsons L, Haque E, Liu H (2004) Subspace clustering for high dimensional data: a review. SIGKDD Explor 6(1):90–105

Vichi M, Rocci R, Kiers HAL (2007) Simultaneous component and clustering models for three-way data: within and between approaches. J Classif 24(1):71–98

Zhang T, Ramakrishnan R, Livny M (1996) Birch: an efficient data clustering method for very large databases. In: Proceedings of the 1996 ACM SIGMOD international conference on management of data, Montreal. ACM, New York, pp 103–114

Part II
Analysis of Data and Models

Bayesian Methods for Conjoint Analysis-Based Predictions: Do We Still Need Latent Classes?

Daniel Baier

Abstract Recently, more and more Bayesian methods have been proposed for modeling heterogeneous preference structures of consumers (see, e.g., Allenby et al., J Mark Res 32:152–162, 1995, 35:384–389, 1998; Baier and Polasek, Stud Classif Data Anal Knowl Organ 22:413–421, 2003; Otter et al., Int J Res Mark 21(3):285–297, 2004). Comparisons have shown that these new methods compete well with the traditional ones where latent classes are used for this purpose (see Ramaswamy and Cohen (2007) Latent class models for conjoint analysis. In: Gustafsson A, Herrmann A, Huber (eds) Conjoint measurement – methods and applications, 4th edn. Springer, Berlin, pp 295–320) for an overview on these traditional methods). This applies especially when the prediction of choices among products is the main objective (e.g. Moore et al., Mark Lett 9(2):195–207, 1998; Andrews et al., J Mark Res 39:479–487, 2002a; 39:87–98, 2002b; Moore, Int J Res Mark 21:299–312, 2004; Karniouchina et al., Eur J Oper Res 19(1):340–348, 2009, with comparative results). However, the question is still open whether this superiority still holds when the latent class approach is combined with the Bayesian one. This paper responds to this question. Bayesian methods with and without latent classes are used for modeling heterogeneous preference structures of consumers and for predicting choices among competing products. The results show a clear superiority of the combined approach over the purely Bayesian one. It seems that we still need latent classes for conjoint analysis-based predictions.

D. Baier (✉)
Chair of Marketing and Innovation Management, Brandenburg University of Technology Cottbus, Postbox 101344, 03013 Cottbus, Germany
e-mail: daniel.baier@tu-cottbus.de

W. Gaul et al. (eds.), *German-Japanese Interchange of Data Analysis Results*,
Studies in Classification, Data Analysis, and Knowledge Organization,
DOI 10.1007/978-3-319-01264-3_9,
© Springer International Publishing Switzerland 2014

103

1 Introduction

Since many years conjoint analysis has proven to be a useful modeling approach when preference structures of consumers w.r.t. attributes and levels of competing products have to be modeled (see, e.g. Green and Rao 1971; Green et al. 2001; Baier and Brusch 2009). Preferential evaluations of sample products (attribute-level-combinations) are collected from sample consumers and for each consumer the relation between attribute-levels and preference values is modeled. Then, these individual models can be used for predicting choices of these consumers in different scenarios. Since in conjoint analysis typically the number of evaluations is low compared to the number of model parameters and many consumers show a similar preference structure, various approaches have been proposed that assume identical model parameters so that the ratio between evaluations and model parameters and – hopefully – the choice predictions using these model parameters can be improved.

Besides approaches that assume the same model parameters across all consumers especially latent class approaches have been proposed for this purpose (see Ramaswamy and Cohen 2007 for an overview on these traditional methods). Here, a division of the market into segments or (latent) classes with homogeneous preference structures is assumed and modeled by identical model parameters within a class. During the modeling step, the class-specific model parameters as well as the number and the size of the classes have to be estimated. Latent Class Metric Conjoint Analysis (shortly: LCMCA, DeSarbo et al. (1992)) is one of the most popular approaches of this kind. In the upper part of Fig. 1 a typical situation is given: The diagrams show a market with three market segments that differ w.r.t. to their preference for "high quality" and for "modern" products. Since the market seems to be clearly segmented, the sharing of evaluations within these segments could lead to an improvement of choice predictions.

Alternatively, recently, Hierarchical Bayesian procedures have been proposed for the same purpose (see, e.g. Allenby et al. 1995, 1998; Lenk et al. 1996). Here, no explicit market segmentation with identical model parameters within the segments is assumed. Instead, a common distribution of the model parameters is postulated for all consumers (first level model), which then is adjusted to individual consumers using their individual evaluations (second level model). Hierarchical Bayes Metric Conjoint Analysis (shortly: HB/MCA, Lenk et al. (1996)) is a popular approach of this kind. In the lower part of Fig. 1 a typical situation is given, where this approach is useful: The diagrams show a market obviously without segments. Consumers differ individually w.r.t. to their preference for "high quality" and for "modern" products, however, they cannot be grouped consistently into homogeneous segments. Market researchers call this situation the "water melon problem" (see, e.g. Sentis and Li 2002): Each dividing up into segments seems to be arbitrarily, so the sharing of evaluations within segments should lead to no improvement of choice predictions. Recently, many comparison studies have shown, that these Hierarchical Bayes approaches seem to compete well with the traditional latent class approaches w.r.t. criteria like model fit or predictive validity (see Table 1 for an overview on comparison studies and their results).

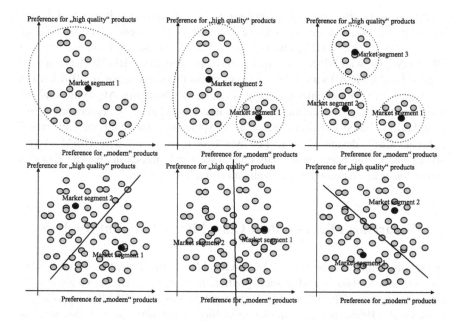

Fig. 1 A market with three market segments (*upper part*) and one without obvious segments (*lower part*); *grey points* indicate individual preferences, *black points* mean preferences when grouping the individuals; the *lines* are used to indicate the allocation of individuals to groups; in the lower – unsegmented – part their exists no obvious grouping

Table 1 Segmentation gains for conjoint analysis-based choice predictions: an overview

Reference	Segm. bases	Criteria	Result
Lenk et al. (1996)	2 real	Pred. validity (FCH, RMSE)	No segm. gains
Allenby et al. (1998)	3 real	Pred. validity (FCH, RMSE)	No segm. gains
Moore et al. (1998)	2 real	Pred. validity (FCH, RMSE)	No segm. gains
Andrews et al. (2002a)	Simulated	Model fit (Pearson, Kendall),Pred. validity (FCH, RMSE)	No sign. diff.
Andrews et al. (2002b)	Simulated	Model fit (Pearson, Kendall),Pred. validity (FCH, RMSE)	No sign. diff.
Gensler (2003)	Simulated	Model fit (Pearson, Kendall),Pred. validity (FCH, RMSE)	No sign. diff.
Moore (2004)	1 real	Pred. validity (FCH, RMSE)	No segm. gains
Karniouchina et al. (2009)	1 real	Pred. validity (FCH, RMSE)	No segm. gains

Across all studies, the assumption of market segments leads to no or only few segmentation gains (i.e. no significant differences w.r.t. model fit or predictive validity) and one could draw the conclusion that we don't need latent classes for conjoint analysis-based choice predictions. However, up to now, it is not clear whether this is also true for a combination of Hierarchical Bayes and Latent Class approaches. For this reason, we compare in this paper a version of such combined approaches, Hierarchical Bayes Latent Class Metric Conjoint Analysis (HB/LCMCA), with HB/MCA, a purely Bayesian one. Since HB/MCA is a special case of HB/LCMCA (with only one latent class) the introduction of HB/LCMCA in chapter "The Randomized Greedy Modularity Clustering Algorithm and the Core Groups Graph Clustering Scheme" suffices. In chapter "Comparison of Two Distribution Valued Dissimilarities and Its Application for Symbolic Clustering" a Monte Carlo design is developed which is used to compare HB/MCA and HB/LCMCA. The paper closes with conclusions and outlook in chapter "Pairwise Data Clustering Accompanied by Validation and Visualisation".

2 Hierarchical Bayes Latent Class Metric Conjoint Analysis

In the following a combination of Hierarchical Bayes and Latent Class approaches for conjoint analysis-based choice prediction is introduced for answering the research question. The HB/LCMCA approach follows the HB/MCA approach in Lenk et al. (1996), but uses similar modeling assumptions as in DeSarbo et al. (1992) for the latent class part of the model and as in Baier and Polasek (2003) for the distributional assumptions. HB/LCMCA contains HB/MCA as a special case (with only one latent class). As in Lenk et al. (1996) the preferential evaluations are modeled as the addition of corresponding partworths (preferential evaluations of attribute-levels).

2.1 The Data, the Model, and the Model Parameters

Let $\mathbf{y}_1, \ldots, \mathbf{y}_n \in \mathbb{R}^m$ describe observed preferential evaluations from n consumers $(i = 1, \ldots, n)$ w.r.t. to m products $(j = 1, \ldots, m)$. y_{ij} denotes the observed preference value of consumer i w.r.t. product j. As an example, these preference values could come from a response scale with values -5 ("I totally dislike this product.") to $+5$ ("I totally like this product."). $\mathbf{X} \in \mathbb{R}^{m \times p}$ denotes the characterization of the m products using p variables. As an example, cars could be characterized by attributes like price, performance, weight, and so on. For estimating the effects of the different attributes on the consumer's preference evaluations, one uses a set of products that reflects possible attribute-levels (e.g. a "low" and a "high" price) in an adequate way, using, e.g., factorial designs w.r.t. to nominal scaled attributes. In this case for $\mathbf{X}$ dummy coded variables are used instead of the original (possibly nominal) attributes.

The observed evaluations are assumed to come from the following model

$$\mathbf{y}_i = \mathbf{X}\boldsymbol{\beta}_i + \boldsymbol{\epsilon}_i, \quad \text{for } i = 1, \ldots, n \quad \text{with } \boldsymbol{\epsilon}_i \sim N(\mathbf{0}, \sigma^2\mathbf{I}) \qquad (1)$$

with $\mathbf{I}$ as the identity matrix, σ^2 as an error variance parameter, and individual part-worths $\boldsymbol{\beta}_1, \ldots, \boldsymbol{\beta}_n$ coming from T latent classes ($t = 1, \ldots, T$) with class-specific partworths $\boldsymbol{\mu}_t \in \mathbb{R}^p$ and class-specific (positive definite) variance/covariance matrices $\mathbf{H}_t \in \mathbb{R}^{p \times p}$:

$$
\boldsymbol{\beta}_i \sim \begin{cases} N(\boldsymbol{\mu}_1, \mathbf{H}_1) & \text{if } C_i = 1, \\ \vdots \\ N(\boldsymbol{\mu}_T, \mathbf{H}_T) & \text{if } C_i = T, \end{cases} \qquad i = 1, \ldots, n. \tag{2}
$$

$\mathbf{C} = (C_1, \ldots, C_n)$ indicates the (latent) classes to which the consumers belong with $C_i \in \{1, \ldots, T\}$, $\boldsymbol{\eta} = (\eta_1, \ldots, \eta_T)$ reflects the (related) size of the classes ($\eta_t = \sum_{i=1}^{n} 1_{\{C_i = t\}} / T$).

2.2 The Bayesian Estimation Procedure

For estimating the model parameters $(\boldsymbol{\eta}, \mathbf{C}, \boldsymbol{\mu}_1, \ldots, \boldsymbol{\mu}_T, \mathbf{H}_1, \ldots, \mathbf{H}_T, \sigma^2)$, Bayesian procedures provide a mathematically tractable way that combines prior information about the model parameters with the likelihood function of the observed data. The result of this combination, the posterior distribution of the model parameters, depends on the modeling assumptions and the assumed prior distributions of the model parameters. It can be derived using iterative Gibbs sampling steps as explained in the following. We use variables with one asterisk ("$*$", e.g., $\mathbf{a}_*$) to denote describing variables of an a priori distribution (prior information) and two asterisks ("$**$", e.g., $\mathbf{a}_{**}$) to denote describing variables of a posterior distribution of the model parameters. Note that the describing variables of the a priori distributions and initial values for the model parameters have to be set before estimation whereas the describing variables of the posterior distributions have derived values allowing iteratively to draw values from the posterior distributions resulting in empirical distributions of all model parameters. We use repeatedly the following five steps:

1. Sample the class indicators $\mathbf{C}$ using the likelihood l of the normal distribution

$$
p(C_i = t | \boldsymbol{\eta}, \boldsymbol{\mu}_1, \ldots, \boldsymbol{\mu}_T, \mathbf{H}_1, \ldots, \mathbf{H}_T, \sigma^2, \mathbf{y}_i) \propto l(\mathbf{y}_i | \mathbf{X}\boldsymbol{\mu}_t, \mathbf{X}\mathbf{H}_t\mathbf{X}' + \sigma^2\mathbf{I})\eta_t
$$

(The consumer is allocated to the class that reflects her/his evaluations best.).
2. Sample the class sizes $\boldsymbol{\eta}$ from

$$
p(\boldsymbol{\eta}|\mathbf{C}) \propto \text{Di}(e_{1**}, \ldots, e_{T**}) \text{ with } e_{1**} = e_{1*} + n_1, \ldots, e_{T*} + n_T, n_t = \sum_{i=1}^{n} 1_{\{C_i = t\}}
$$

($\text{Di}(e_1, \ldots, e_T)$ represents the Dirichlet distribution with concentration variables $e_1, \ldots, e_T$. The variables of the a priori distribution are set to 1: $e_{t*} = 1 \ \forall \ t$.).

3. Sample the class-specific partworths $\boldsymbol{\mu}_1, \ldots, \boldsymbol{\mu}_T$ from

$$p((\boldsymbol{\mu}_1', \ldots, \boldsymbol{\mu}_T')'|\mathbf{C}, \mathbf{H}_1, \ldots, \mathbf{H}_T, \sigma^2, \mathbf{y}_1, \ldots, \mathbf{y}_n) \propto N(\mathbf{a}_{**}, \mathbf{A}_{**})$$

$$\text{with } \mathbf{Z}_i = (\mathbf{X}1_{\{C_i=1\}}, \ldots, \mathbf{X}1_{\{C_i=T\}}), \mathbf{V}_i = \mathbf{X}(\mathbf{H}_{C_i}^{-1})\mathbf{X}' + \sigma^2\mathbf{I},$$

$$\mathbf{A}_{**} = (\sum_{i=1}^{n} \mathbf{Z}_i'\mathbf{V}_i^{-1}\mathbf{Z}_i + \mathbf{A}_*^{-1})^{-1}, \quad \mathbf{a}_{**} = \mathbf{A}_{**}(\sum_{i=1}^{n} \mathbf{Z}_i'\mathbf{V}_i^{-1}\mathbf{y}_i + \mathbf{A}_*^{-1}\mathbf{a}_*)$$

(Due to known problems with slow convergence, the class-specific partworths are sampled simultaneously. The class-specific partworths are stacked, $\mathbf{a}_{**}$ and $\mathbf{A}_{**}$ are the mean and the blocked variance/covariance matrix of the corresponding posterior distribution. The variables of the a priori distribution, $\mathbf{a}_*$ and $\mathbf{A}_*$, are set to be non-informative, alternatively, they could be used as in Baier and Polasek (2003) to constrain the partworths. The $\mathbf{Z}_i$ and $\mathbf{V}_i$ matrices are used to allocate the individual evaluations to the corresponding class.).

4. Sample the individual partworths $\boldsymbol{\beta}_1, \ldots, \boldsymbol{\beta}_n$ from

$$p(\boldsymbol{\beta}_1, \ldots, \boldsymbol{\beta}_n|\mathbf{C}, \boldsymbol{\mu}_1, \ldots, \boldsymbol{\mu}_T, \mathbf{H}_1, \ldots, \mathbf{H}_T, \sigma^2, \mathbf{y}_1, \ldots, \mathbf{y}_n) \text{ using } \boldsymbol{\beta}_i \sim N(\mathbf{b}_{i**}, \mathbf{B}_{i**})$$

$$\text{with } \mathbf{B}_{i**} = (\mathbf{X}'\mathbf{X}/\sigma^2 + \mathbf{H}_{C_i}^{-1})^{-1} \text{ and } \mathbf{b}_{i**} = \boldsymbol{\mu}_{C_i} + \mathbf{B}_{i**}\mathbf{X}'\mathbf{y}_i/\sigma^2 + \mathbf{H}_{C_i}^{-1}\boldsymbol{\mu}_{C_i}.$$

(The posterior distribution of the partworths for individual i with describing variables $\mathbf{b}_{i**}$ and $\mathbf{B}_{i**}$ combines the information from the corresponding class-specific partworths with the observed preferential evaluations of individual i.)

5. Sample the variance/covariance model parameters $\mathbf{H}_1, \ldots, \mathbf{H}_T, \sigma^2$ from

$$p(\mathbf{H}_1, \ldots, \mathbf{H}_T, \sigma^2|\boldsymbol{\beta}_1, \ldots, \boldsymbol{\beta}_n, \mathbf{C}, \boldsymbol{\mu}_1, \ldots, \boldsymbol{\mu}_T, \mathbf{y}_1, \ldots, \mathbf{y}_n) \text{ using}$$

$$\mathbf{H}_t \sim IW(w_{t**}, \mathbf{W}_{t**}) \text{ with}$$

$$w_{t**} = w_{t*} + 0.5\sum_{i=1}^{n} 1_{\{C_i=t\}}, \mathbf{W}_{t**} = \mathbf{W}_{t*} + 0.5\sum_{i=1}^{n}(\boldsymbol{\beta}_i - \boldsymbol{\mu}_t)(\boldsymbol{\beta}_i - \boldsymbol{\mu}_t)'1_{\{C_i=t\}} \text{ and}$$

$$\sigma^2 \sim IG(g_{**}, G_{**}) \text{ with } g_{**} = g_* + \frac{nm}{2}, G_{**} = G_* + \frac{1}{2}\sum_{i=1}^{n}(\mathbf{X}\boldsymbol{\beta}_i - \mathbf{y}_i)'(\mathbf{X}\boldsymbol{\beta}_i - \mathbf{y}_i).$$

(IW stands for the Inverse Wishart distribution, IG for the Inverse Gamma distribution. Both distributions are used to model the a priori and the posterior distributions of the variance/covariance model parameters. We use similar settings for the a priori distributions as in Baier and Polasek (2003).)

As usual in Bayesian research, the posterior distribution of the model parameters are empirical distributions which collect the draws of the iterative Gibbs steps. Each empirical distribution consists typically of 1,000–2,000 draws, the "first" draws (e.g. the first 200 draws) are typically discarded due to the need of a so-called "burn-in phase" during estimation.

When latent classes have to be modeled in Bayesian research, often the so-called "relabeling problem" occurs: From a statistical point of view the "labels" of the classes (their number $1,\ldots,T$) provide no information. For one draw of all model parameters, changing the numbers of two or more classes makes no difference ("unidentifiability problem"). However, during the iterative process over 1,000 or more draws, such changes (due to algorithmic indeterminacy) lead to bad results w.r.t. the empirical distributions. Therefore, usually, in step 2 a relabeling is enforced that – after drawing the segment sizes – ensures that the class 1 has the smallest size, 2 the second smallest and so on. Alternatively, the relabeling could take place in step 3 w.r.t. class-specific partworths by ensuring that the importance of, e.g., attribute 1 is highest for class 1, second highest for class 2, and so on.

2.3 Model Fit and Predictive Validity

Once the posterior distribution of the parameters is available one can control model fit or predictive validity in various ways. So, w.r.t. model fit, the preferential evaluations w.r.t. to the estimation sample of evaluations could be compared with the corresponding predictions using Pearson's correlation coefficient. W.r.t. predictive validity one uses the possibility that the model can also be used to predict preferential evaluations w.r.t. modified sets of products (scenarios) by changing m and $\mathbf{X}$ accordingly. One collects additional preferential evaluations w.r.t. to so-called hold-out products and compares this evaluations with predictions of the model using criteria like the Root Mean Squared Error (RMSE) which stands for the deviation between the observed and predicted preferential evaluations or the first choice which stands for the percentage of predictions where the "best" holdout product w.r.t to the observed and predicted evaluations is the same.

3 Monte Carlo Comparison of HB/MCA and HB/LCMCA

In order to decide whether one still needs latent classes for conjoint analysis-based choice predictions a comprehensive Monte Carlo analysis was performed to compare the purely Bayesian approach (HB/MCA) with the combination of the Bayesian approach and latent class modeling. One should keep in mind that HB/MCA is the HB/LCMCA version with only one latent class ($T = 1$), so, w.r.t. model fit there should be a superiority of the combined over the purely Bayesian approach. However, the question is, whether this also holds w.r.t. predictive validity.

3.1 Design of the Monte Carlo Study

In total, 1,350 datasets were generated, using 50 replications w.r.t. 3 dataset generation factors with 3 possible levels each (forming $3 \times 3 \times 3 \times 50 = 1,350$ datasets). Each generated dataset describes a conjoint experiment for estimating the preferences of 300 consumers w.r.t to products characterized by 8 two-level attributes. The simulated conjoint task for each consumer was to evaluate a set of 16 products whose dummy coded descriptions w.r.t. the 8 two-level attributes were generated using a Plackett and Burman (1946) factorial design (with 16 rows and 8 columns). Also, a set of 8 additional products was used to generate additional preferential evaluations from each consumer for checking the predictive validity. The first 16 products form the estimation set, the last 8 products the holdout set of products.

A "true" preference structure of the consumers was assumed that could come – according to the first dataset generation factor ("Heterogeneity between segments") – from a market with one, two, or three segments. The market with only one segment is used as a proxy for an unsegmented market, the markets with two or three segments as proxies for segmented markets. As in other simulation studies, the means of the "true" segment-specific partworths were randomly drawn from the $[-1, 1]$ uniform distribution. All in all the following three dataset generation factors were used:

- **Heterogeneity between segments (unsegmented or not segmented market):** For a third of the datasets (level "low" for factor "heterogeneity between segments"), it was assumed that there is no segment-specific preference structure, i.e. all "true" individual partworths are drawn from one (normal) distribution (one market segment). For the other datasets (levels "medium" and "high"), it was assumed that there is a segment-specific preference structure, i.e. all "true" individual partworths are drawn from two ("medium") or three ("high") different (normal) distributions (two or three market segments). The size of these market segments was predefined as 100 % (in the case of one market segment, 300 consumers), 50 and 50 % in the case of two market segments (each segment contains 150 consumers) resp. 50, 30 and 20 % in the case of three market segments (containing 150, 90 and 60 consumers).
- **Heterogeneity within segments (segment-specific distributions of individual partworths):** For all datasets it was assumed that the individual partworths are drawn from normal distributions around the mean of their corresponding segment-specific partworths (drawn from a uniform distribution as described above). The variance/covariance matrix of these normal distributions was assumed to be diagonal with identical values σ^2 in the diagonal. For a third of the datasets these diagonal values (and consequently the heterogeneity within segments) were assumed to be "low ($\sigma = 0.1$)", for another third "medium ($\sigma = 0.25$)", and for another third "high ($\sigma = 0.5$)".
- **Disturbance (additive preference value error in data collection):** Additionally, as in other studies, a measurement error was introduced for the simulated

Table 2 Model fit across the datasets in the Monte Carlo analysis

Factor	Level	HB/MCA		HB/LCMCA	
		Corr($\mathbf{y}_i$)	Corr($\boldsymbol{\beta}_i$)	Corr($\mathbf{y}_i$)	Corr($\boldsymbol{\beta}_i$)
Heterogeneity	Low (450 datasets)	**0.744**	**0.954**	0.742	0.951
Between	Medium (450 datasets)	0.535	0.698	**0.685*****	**0.817*****
Components	High (450 datasets)	0.456	0.576	**0.655*****	**0.767*****
Heterogeneity	Low (450 datasets)	0.594	0.763	**0.724*****	**0.872*****
Within	Medium (450 datasets)	0.612	0.784	**0.731*****	**0.884*****
Components	High (450 datasets)	0.529	0.682	**0.627*****	**0.779*****
Disturbance	Low (450 datasets)	0.746	0.752	**0.930*****	**0.936*****
(Data error)	Medium (450 datasets)	0.686	0.759	**0.847*****	**0.925*****
	High (450 datasets)	0.303	**0.718***	0.305	0.674
Total (1,350 datasets)		0.578	0.743	**0.694*****	**0.845*****

* Significant differences at $\alpha = 0.1$; **at $\alpha = 0.01$; ***at $\alpha = 0.001$

data collection step. The calculated preference values for each product using the generated "true" individual partworths were superimposed by a normally distributed additive error (see model formulation in Sect. 2.1) with a "low ($\sigma = 0.4$)", "medium ($\sigma = 1$)" or "high ($\sigma = 2$)" standard deviation.

For each possible factor-level-combination – a total of $3 \times 3 \times 3 = 27$ combinations was possible – the dataset generation was repeated 50 times (full factorial design with 50 repetitions). As a result each dataset comprised conjoint evaluations from 300 consumers with respect to 16 products for estimation (using – as above mentioned – a Plackett and Burman (1946) factorial design) and 8 randomly generated holdout products for checking the predictive validity. It should be mentioned that – besides transforming the generated preferential evaluations into a Likert scale – the dataset generation process reflects the model formulation quite good (as usual, see the simulation studies in Table 1).

The HB/MCA and HB/LCMCA procedures were used with non-informative priors in order not to distort the estimation results by information outside the available collected data w.r.t. the 16 products. The number of segments (T) was predefined according to the HB/MCA ($T = 1$) or HB/LCMCA ($T = 2, 3$) procedure. For all estimations, 1,000 Gibbs iterations with 200 burn-ins proved to be sufficient for convergence. For HB/LCMCA relabeling w.r.t. to the class size (label order equals size order) was used.

3.2 Results w.r.t. Model Fit

For checking the model fit, mean Pearson correlation coefficients between true and estimated individual preference values for products (Corr($\mathbf{y}_i$)) as well as mean Pearson correlation coefficients between true and estimated individual partworths (Corr($\boldsymbol{\beta}_i$)) were calculated. Table 2 shows aggregated results (mean values w.r.t. to

Table 3 Predictive validity across the datasets in the Monte Carlo analysis

		HB/MCA		HB/LCMCA	
Factor	Level	First choice	RMSE	First choice	RMSE
Heterogeneity	Low (450 datasets)	**0.775**	**0.060**	0.770	0.062
Between	Medium (450 datasets)	0.567	0.121	**0.702*****	**0.066*****
Components	High (450 datasets)	0.491	0.132	**0.673*****	**0.063*****
Heterogeneity	Low (450 datasets)	0.640	0.099	**0.757*****	**0.056*****
Within	Medium (450 datasets)	0.631	0.099	**0.761*****	**0.052*****
Components	High (450 datasets)	0.563	0.116	**0.628*****	**0.083*****
Disturbance	Low (450 datasets)	0.687	0.087	**0.870*****	**0.029*****
(Data error)	Medium (450 datasets)	0.677	0.088	**0.798*****	**0.041*****
	High (450 datasets)	0.469	0.139	**0.478**	**0.120**
Total (1,350 datasets)		0.611	0.105	**0.715*****	**0.063*****

* Significant differences at $\alpha = 0.1$; **at $\alpha = 0.01$; ***at $\alpha = 0.001$

the Pearson correlation coefficients) across all datasets with one factor-level fixed $(3 \times 3 \times 50 = 450$ datasets) and across all datasets $(3 \times 3 \times 3 \times 50 = 1,350$ datasets).

For each factor-level combination of the Monte Carlo analysis these values were calculated and compared between HB/MCA and HB/LCMCA. The results are convincing: If a segment-specific structure is in the data, the segment-free HB/MCA is outperformed by the segment-specific HB/LCMCA procedure. Overall the superiority can clearly be seen.

3.3 Results w.r.t. Predictive Validity

In a similar way, the predictive validity was checked. For the eight holdout products and each consumer, preference values were calculated from the estimated individual partworths and compared to the preference values that were derived from the "true" partworths. As criteria for the comparison the so-called first choice hit rate (first choice) and mean root mean squared error (RMSE) were calculated. First choice hit indicates for a consumer whether her/his preference values from the estimated and from the "true" partworths are maximum for the same holdout product, the first choice hit rate is the share of consumers where a first choice hit occurs. RMSE compares also the preference values from the estimated and from the "true" partworths but more according to their absolute values.

Table 3 shows (again) aggregated results (mean values w.r.t. to the first choice hit rate and RMSE) across all datasets with one factor-level fixed $(3 \times 3 \times 50 = 450$ datasets) and across all datasets $(3 \times 3 \times 3 \times 50 = 1,350$ datasets). Again, the results are convincing: If a segment-specific structure is in the data, the segment-free HB/MCA is outperformed by the segment-specific HB/LCMCA procedure. Overall the superiority of the combined approach can clearly be seen.

4   Conclusions and Outlook

The comparison in this paper clearly shows that we still need latent classes for conjoint analysis-based predictions even if we use Bayesian procedures for parameter estimation. HB/LCMCA was clearly superior to HB/MCA w.r.t. model fit and predictive validity, especially in cases when markets are segmented. However, these results are only based on a rather small number of datasets (1,350 datasets) generated synthetically and therefore no real data. More research in this field needs to be done, especially with a larger set of conjoint data from real markets.

References

Allenby GM, Arora N, Ginter JL (1995) Incorporating prior knowledge into the analysis of conjoint studies. J Mark Res 32:152–162

Allenby GM, Arora N, Ginter JL (1998) On the heterogeneity of demand. J Mark Res 35:384–389

Andrews RL, Ainslie A, Currim IS (2002a) An empirical comparison of logit choice models with discrete versus continuous representations of heterogeneity. J Mark Res 39:479–487

Andrews RL, Ainslie A, Currim IS (2002b) Hierarchical Bayes versus finite mixture conjoint analysis models: a comparison of fit, prediction, and partworth recovery. J Mark Res 39:87–98

Baier D, Brusch M (2009) Conjointanalyse: Methoden – Anwendungen – Praxisbeispiele. Springer, Berlin

Baier D, Polasek W (2003) Market simulation using Bayesian procedures in conjoint analysis. Stud Classif Data Anal Knowl Organ 22:413–421

DeSarbo WS, Wedel M, Vriens M, Ramaswamy V (1992) Latent class metric conjoint analysis. Mark Lett 3(3):273–288

Gensler S (2003) Heterogenität in der Präferenzanalyse. Ein Vergleich von hierarchischen Bayes-Modellen und Finite-Mixture-Modellen. Gabler, Wiesbaden

Green PE, Rao VR (1971) Conjoint measurement for quantifying judgmental data. J Mark Res 8(3):355–363

Green PE, Krieger AM, Wind Y (2001) Thirty years of conjoint analysis: reflections and prospects. Interface 31(3):S56–S73

Karniouchina EV, Moore WL, Rhee BVD, Vermad R (2009) Issues in the use of ratings-based versus choice-based conjoint analysis in operations management research. Eur J Oper Res 19(1):340–348

Lenk PJ, DeSarbo WS, Green PE, Young MR (1996) Hierarchical Bayes conjoint analysis: recovery of partworth heterogeneity from reduced experimental designs. Mark Sci 15(2):173–191

Moore WM (2004) A cross-validity comparison of rating-based and choice-based conjoint analysis models. Int J Res Mark 21:299–312

Moore WL, Gray-Lee J, Louviere JJ (1998) A cross-validity comparison of conjoint analysis and choice models at different levels of aggregation. Mark Lett 9(2):195–207

Otter T, Tüchler R, Frühwirth-Schnatter S (2004) Capturing consumer heterogeneity in metric conjoint analysis using Bayesian mixture models. Int J Res Mark 21(3):285–297

Plackett RL, Burman JP (1946) The design of optimum multifactorial experiments. Biometrika 33:305–325

Ramaswamy V, Cohen SH (2007) Latent class models for conjoint analysis. In: Gustafsson A, Herrmann A, Huber (eds) Conjoint measurement – methods and applications, 4th edn. Springer, Berlin, pp 295–320

Sentis K, Li L (2002) One size fits all or custom tailored: which HB fits better? In: Proceedings of the sawtooth software conference, Sequim, pp 167–175, Sept 2001

Non-additive Utility Functions: Choquet Integral Versus Weighted DNF Formulas

Eyke Hüllermeier and Ingo Schmitt

Abstract In the context of conjoint analysis, a consumer's purchase preferences can be modeled by means of a utility function that maps an attribute vector describing a product to a real number reflecting the preference for that product. Since simple additive utility functions are not able to capture interactions between different attributes, several types of non-additive functions have been proposed in recent years. In this paper, we compare two such model classes, namely the (discrete) Choquet integral and weighted DNF formulas as used in a logic-based query language called CQQL. Although both approaches have been developed independently of each other in different fields (decision analysis and information retrieval), they are actually quite similar and share several commonalities. By developing a conceptual link between the two approaches, we provide new insights that help to decide which of the two alternatives is to be preferred under what conditions.

1 Introduction

The modeling of a decision maker's preferences is of major concern in a multitude of research areas, ranging from more traditional ones like economics and operations research to modern fields like artificial intelligence and information retrieval (Domshlak et al. 2011). Apart from modeling and reasoning about preferences, the problem of learning preference models for predictive purposes has

E. Hüllermeier

Department of Mathematics and Computer Science, University of Marburg, Marburg, Germany
e-mail: eyke@mathematik.uni-marburg.de

I. Schmitt (✉)

Institute of Computer Science, Technical University of Cottbus, Cottbus, Germany
e-mail: schmitt@tu-cottbus.de

W. Gaul et al. (eds.), *German-Japanese Interchange of Data Analysis Results*,
Studies in Classification, Data Analysis, and Knowledge Organization,
DOI 10.1007/978-3-319-01264-3_10,
© Springer International Publishing Switzerland 2014

attracted increasing attention in recent years (Fürnkranz and Hüllermeier 2011). Take conjoint analysis as an example, where the goal is to predict the purchase decisions of consumers (Baier and Brusch 2009; Green and Rao 1971). To this end, a consumer's preferences are typically expressed in terms of a *utility function*, which maps an attribute vector describing a product to a real number reflecting the preference for that product.

More formally, consider a set O of objects, and assume that all objects $o \in O$ share the same set of attributes $C = \{c_1, \ldots, c_m\}$; each attribute c_i typically corresponds to a "criterion" reflecting a certain aspect of a decision situation (e.g., the price of a product to be purchased). For an object o, the function $f_o : C \rightarrow \mathbf{R}_+$ assigns real values to attributes. Moreover, a utility function $U \in \mathcal{U}$, where $\mathcal{U}$ is an underlying (parameterized) class of functions, aggregates these values and combines them into a single utility score.[1] Thus, it maps each vector $(x_1, \ldots, x_m) = (f_o(c_1), \ldots, f_o(c_m)) \in \mathbf{R}_+^m$ to a non-negative real number. By convention, higher utility scores indicate higher preference.

The problem of learning a utility function from observed preference data (e.g., consumer preferences revealed through purchase decisions) essentially comes down to estimating the parameters of that function. With respect to the generalization performance of this model, i.e., its ability to accurately predict the utility of objects not seen so far, the "capacity" of the model class $\mathcal{U}$ is of utmost importance. On the one side, $\mathcal{U}$ must be expressive enough, so as to allow for approximating the target (e.g., the utility function of a consumer) sufficiently well. On the other hand, it must not be overly flexible, so as to avoid poor generalization performance caused by an overfitting of the training data.

In many cases, simple additive utility functions of the form $U(x_1, \ldots, x_m) = w_0 + \sum_i w_i x_i$ are not expressive enough. In particular, they are not able to capture any interactions between the attributes c_i. Therefore, several types of non-additive functions have been proposed in recent years. In this paper, we compare two such functions, namely the (discrete) Choquet integral (Choquet 1954) and weighted DNF formulas expressed in a logic-based query language called CQQL (Schmitt 2008). Although both approaches have been developed independently of each other in different fields (decision analysis and information retrieval), they are actually quite similar and share several commonalities; besides, both of them have already been used in a machine learning context (Tehrani et al. 2012; Baier and Schmitt 2012).

Specifically, we will show that the Choquet integral and the weighted DNF formulas can be seen, respectively, as the set-based and the logic-based version of almost the same mathematical structure. Our contribution is the discussion of similarities and differences of the two approaches, as well as the establishment of a conceptual link between them. This link provides new insights and helps to decide which of the two alternatives is to be preferred under what conditions.

---

[1]The function class $\mathcal{U}$ is typically chosen so as to guarantee reasonable properties (rationality axioms) of the aggregation process.

2 Discrete Choquet Integral

Let $C = \{c_1, \ldots, c_m\}$ be a finite set of attributes or criteria as introduced above, and $\mu(\cdot)$ a measure $2^C \to [0, 1]$. For each $A \subseteq C$, we interpret $\mu(A)$ as the *weight* or, say, the *importance* of the set of criteria A. A standard assumption on a measure $\mu(\cdot)$, which is, for example, at the core of probability theory, is additivity: $\mu(A \cup B) = \mu(A) + \mu(B)$ for all $A, B \subseteq C$ such that $A \cap B = \emptyset$. Unfortunately, additive measures cannot model any kind of interaction between criteria, since the increase of the measure due to the addition of a subset of attributes $B \subseteq C$ is always the same and only depends on B itself. More specifically, adding B to a set of criteria A (not intersecting with B) always increases the weight of the measure by $\mu(B)$, regardless of A.

Non-additive measures, also called capacities or fuzzy measures, are simply normalized and monotone (Sugeno 1974): $\mu(\emptyset) = 0, \mu(C) = 1$, and $\mu(A) \le \mu(B)$ for all $A \subseteq B \subseteq C$. A useful representation of non-additive measures, that we shall explore later on, is in terms of the *Möbius transform* (Rota 1964):

$$\mu(B) = \sum_{A \subseteq B} m_\mu(A) \tag{1}$$

for all $B \subseteq C$, where the Möbius transform m_μ of the measure μ is defined as follows:

$$m_\mu(A) = \sum_{B \subseteq A} (-1)^{|A|-|B|} \mu(B). \tag{2}$$

The value $m_\mu(A)$ can be interpreted as the weight that is *exclusively* allocated to A, instead of being indirectly connected with A through the interaction with other subsets.

A measure μ is said to be k-additive, if k is the smallest integer such that $m_\mu(A) = 0$ for all $A \subseteq C$ with $|A| > k$. This property is interesting for several reasons. First, as can be seen from (1), it means that a measure μ can formally be specified by significantly fewer than 2^m values which are needed in the general case. Second, k-additivity is also interesting from a semantic point of view: This property simply means that there are no interaction effects for subsets $A \subseteq C$ whose cardinality exceeds k.

While $w_i = \mu(\{c_i\})$ is a natural quantification of the *importance* of a criterion c_i in the case of an additive measure (utility function), measuring the importance of a criterion becomes obviously more involved when μ is non-additive. Besides, one may then also be interested in a measure of *interaction* between the criteria, either pairwise or even of a higher order. In the literature, measures of that kind have been proposed, both for the importance of single (Shapley 1953) as well as the interaction between several criteria (Murofushi and Soneda 1993).

Given a fuzzy measure μ on C, the *Shapley value* (or importance index) of c_i is defined as a kind of average increase in importance due to adding c_i to another subset $A \subset C$ (Shapley 1953):

$$\varphi(c_i) = \sum_{A \subseteq C \setminus \{c_i\}} \frac{1}{m \binom{m-1}{|A|}} \Big(\mu(A \cup \{c_i\}) - \mu(A)\Big). \qquad (3)$$

The Shapley value of μ is the vector $\varphi(\mu) = (\varphi(c_1), \ldots, \varphi(c_m))$. One can show that $0 \le \varphi(c_i) \le 1$ and $\sum_{i=1}^{m} \varphi(c_i) = 1$. Thus, $\varphi(c_i)$ is a measure of the *relative* importance of c_i. Obviously, $\varphi(c_i) = \mu(\{c_i\})$ if μ is additive.

The *interaction index* between criteria c_i and c_j, as proposed by Murofushi and Soneda (1993), is defined as follows:

$$I_{i,j} = \sum_{A \subseteq C \setminus \{c_i, c_j\}} \frac{\mu(A \cup \{c_i, c_j\}) - \mu(A \cup \{c_i\}) - \mu(A \cup \{c_j\}) + \mu(A)}{(m-1) \binom{m-2}{|A|}}.$$

This index ranges between -1 and 1 and indicates a positive (negative) interaction between criteria c_i and c_j if $I_{i,j} > 0$ ($I_{i,j} < 0$). The definition of interaction can also be extended to more than two criteria (Grabisch 1997).

For a given object suppose that $f : C \to [0, 1]$ assigns a normalized *value* from the unit interval to each criterion c_i. An important question, then, is how to *aggregate* the evaluations of individual criteria, i.e., the values $f(c_i)$, into an overall evaluation, in which the criteria are properly weighted according to the measure μ. Mathematically, this overall evaluation can be considered as an integral $\mathscr{C}_\mu(f)$ of the function f with respect to the measure μ. Indeed, if μ is an additive measure, the standard integral just corresponds to the *weighted mean*

$$\mathscr{C}_\mu(f) = \sum_{i=1}^{m} w_i \cdot f(c_i) = \sum_{i=1}^{m} \mu(\{c_i\}) \cdot f(c_i), \qquad (4)$$

which is a natural aggregation operator in this case. A non-trivial question, however, is how to generalize (4) in the case where μ is non-additive. This question is answered by the Choquet integral, which, in the discrete case, is formally defined as follows:

$$\mathscr{C}_\mu(f) = \sum_{i=1}^{m} \big(f(c_{(i)}) - f(c_{(i-1)})\big) \cdot \mu(A_{(i)}) \ ,$$

where $(\cdot)$ is a permutation of the index set $\{1, \ldots, m\}$ that sorts the criteria in increasing order of their evaluation, i.e., such that $0 \le f(c_{(1)}) \le f(c_{(2)}) \le \cdots \le f(c_{(m)})$ (and $f(c_{(0)}) = 0$ by definition), and $A_{(i)} = \{c_{(i)}, \ldots, c_{(m)}\}$. In terms of the Möbius transform $m = m_\mu$ of μ, the Choquet integral can also be expressed as follows:

$$C_\mu(f) = \sum_{T \subseteq C} m_\mu(T) \times \min_{c_i \in T} f(c_i) \tag{5}$$

3   Weighted DNF Formula

The Commuting Quantum Query Language (CQQL, Schmitt 2008) is a database query language that extends the relational domain calculus (Codd 1971; Maier 1983) by the notion of proximity. By using concepts from quantum logic (von Neumann 1932), CQQL reconciles the logical operators *conjunction, disjunction,* and *negation* with score values from $[0, 1] \subseteq \mathbf{R}$. Schmitt (2008) shows that these logical operators in CQQL obey the rules of a Boolean algebra. A score value results from evaluating a proximity predicate $f^p \in [0, 1]$ against a value $f(c_i)$ of an object attribute c_i.

Example 1. The proximity predicate *price_is_high* on the price of a product is given by $f^p(f(c_{price})) = e^{-|f(c_{prize})-maxp|}$, where *maxp* is the maximal possible price. This predicate returns a high value close to 1 if the price of a product is high.

For simplicity, we will write $f^p(c)$ instead of $f^p(f(c))$.

The syntax of a CQQL query is almost the same as the syntax of the relational domain calculus.[2] CQQL offers weighted versions of conjunction and disjunction, which allows for controlling the influence of their operands by using weights from $[0, 1]$. An operand with weight 0 has no effect, whereas one with a weight 1 equals the unweighted case.

From the laws of logic, we know that every Boolean algebra formula over an attribute set $C = \{c_1, \ldots, c_m\}$ can be expressed in full disjunctive normal form (DNF). Thus, every formula is bidirectly associated to a subset of 2^m minterms. Our weighted DNF approach is to assign a weight λ out of $[0, 1]$ to every DNF minterm T (Schmitt and Zellhöfer 2012):

$$\bigvee_{T \subseteq C}^{\lambda(T)} \left(\bigwedge_{c_i \in T} f^p(c_i) \wedge \bigwedge_{c_i \in C \setminus T} \neg f^p(c_i) \right) = \bigvee_{T \subseteq C} \left(\lambda(T) \wedge \bigwedge_{c_i \in T} f^p(c_i) \wedge \bigwedge_{c_i \in C \setminus T} \neg f^p(c_i) \right).$$

The weights $\lambda(T)$ can be seen as the parameters of the weighted DNF model class, and a concrete model is identified by a mapping λ of all 2^m minterms into $[0, 1]^{2^m}$ (Zellhöfer and Schmitt 2010). Due to the full disjunctive normal form, comparing any two different minterms at least one predicate is negated in one and

---

[2]In the following we will neglect quantifiers and functions.

not negated in the second minterm. Because of exclusive disjunctions, the arithmetic
CQQL evaluation reduces to a simple weighted sum:

$$wDNF_\lambda(f^p) = \sum_{T \subseteq C} \lambda(T) \times \prod_{c_i \in T} f^p(c_i) \times \prod_{c_i \in C \setminus T} (1 - f^p(c_i)) \qquad (6)$$

Applying the Möbius transform $(m_\lambda(A) = \sum_{B \subseteq A} (-1)^{|A|-|B|} \lambda(B))$ on the weights
yields:

$$wDNF_\lambda(f^p) = \sum_{T \subseteq C} m_\lambda(T) \times \prod_{c_i \in T} f^p(c_i). \qquad (7)$$

4   Choquet Integral Versus Weighted DNF Formulas

In this section, we examine commonalities and differences between the Choquet
integral and the weighted DNF approach. Comparing (5) with (7), one recognizes
the same mathematical structure. Both formulas are summing over all subsets of
attributes, and every subset is weighted. However, the formulas differ in their last
term: $\min_{c_i \in T} f(c_i)$ versus $\prod_{c_i \in T} f^p(c_i)$. From the point of view of fuzzy logic,
both terms are conceptually strongly related, since both the min and the product
operator are triangular norms, that is, generalized logical conjunctions (Klement
et al. 2002).

There are, however, also important differences between (5) with (7). First,
although both terms $f(c_i)$ and $f^p(c_i)$ assume values in the unit interval, they are
interpreted in different ways: In the case of the Choquet integral, $x_i = f(c_i)$
is considered as an individual utility degree, that is, the degree to which the
i-th criterion c_i is satisfied. Thus, x_i is a "the higher—the better" attribute; ideally,
$x_i = 1$, whereas $x_i = 0$ corresponds to the least preferred situation. In CQQL,
the meaning of $x_i = f^p(c_i)$ is more general. In fact, f^p is more neutral and
merely expresses proximity, while not yet committing to any sort of preference.
In particular, there is no clear direction of preference, i.e., x_i is not necessarily a
"the higher—the better" attribute.

The second important difference, which is closely connected to the first one,
concerns the monotonicity of the measure μ in the Choquet integral, and the
constraints implied by this property on the Moebius transform m_μ in (5). Given
the interpretation of $\mu(A)$ as the *overall* importance of the attribute subset $A \subseteq C$,
the monotonicity property is rather natural. Correspondingly, the Choquet integral
is a *monotone* aggregation operator: Increasing (decreasing) the values x_i of the
criteria $c_i \in A$ can only increase (decrease) the overall utility, and the corresponding
degree of increase (decrease) depends on the importance of A and the way in which
this subset of criteria interacts with other criteria.

In a weighted DNF formula, there is no monotonicity of that kind. Instead, an increase of $x_i = f^p(c_i)$ may lead to both, an increase or a decrease of $wDNF_\lambda(f^p)$. In fact, the effect will depend on the context, that is, the values of all other attributes. Thus, by relaxing the monotonicity constraint, the weighted DNF can in a sense be seen as a generalization of the Choquet integral. As such, it is able to model dependencies that cannot be captured by the latter. As a striking example, consider the XOR combination of the values x_i (i.e., the overall utility is 1 if $|\{i \mid x_i = 1\}|$ is odd and 0 otherwise). Here, the effect of increasing a single x_i from 0 to 1 can be positive or negative, depending on the values of the other attributes (for example, a dish may become more delicious by adding either turmeric or ginger, but not both). While this dependency (even in a weighted version) can be modeled by a weighted DNF formula, it obviously violates the monotonicity properties of the Choquet integral.

From the discussion so far, it is clear that, from a modeling point of view, a weighted DNF formula is even more flexible than the Choquet integral. This observation can also be stated more formally, for example in terms of the so-called Vapnik–Chervonenkis (VC) dimension (Vapnik 1998). The VC dimension is a measure of the flexibility (capacity) of a model class. It is a key notion in statistical learning theory and plays an important role in estimating the generalization performance of a learning method. Recently, it has been shown that the VC dimension of the Choquet integral, when being used as a threshold classifier, grows asymptotically at least as fast as $2^m / \sqrt{m}$, where m is the number of attributes (Hüllermeier and Tehrani 2012). For the weighted DNF formula, it is not difficult to show that the VC dimension is 2^m. Due to the monotonicity constraint, the VC dimension of the Choquet integral is reduced, albeit only by a (surprisingly small) factor of $\sqrt{m}$. Nevertheless, both model classes are extremely flexible. Consequently, successful learning will only become possible through proper regularization techniques.

In the case of the Choquet integral, this can be accomplished by restricting the underlying measure μ to k-additive measures (cf. Sect. 2). This is simply realized by setting some of the coefficients in (5) to 0 (namely those $m_\mu(A)$ for $|A| > k$). Analogously, a restriction towards a monotone and k-additive measure can be realized in the arithmetic weighted DNF formula by restricting the λ-weights in (7). This can be accomplished quite easily by adding some linear inequalities to the linear problem. However, the so far open question arises, how theses restrictions can be expressed within the logic-based CQQL formalism. It can be easily shown that by means of so called *connected weights* the CQQL formalism is expressive enough to represent an additive measure.

As one of the nice features of a monotone measure, we already mentioned the existence of importance and interaction indices that can be derived from the measure μ. Thus, despite the non-additivity of the measure and, coming along with this, the non-linearity (of the utility function), the influence of single attributes and the interaction between groups of attributes can be characterized in a formally sound way.

5 Summary and Outlook

Our comparison of the (discrete) Choquet integral and the weighted DNF formulas as non-linear generalizations of additive utility models has shown that both approaches are indeed closely related. The most important difference between them concerns the monotonicity property of the Choquet integral. On the one side, this property restricts the class of utility functions that can be modeled, on the other side, it arguably facilitates the interpretation of a model, notably through the derivation of indices characterizing the underlying non-additive measure (such as Shapley value and interaction index). Of course, the weighted DNF formula (7) can be easily restricted to monotonicity and k-additivity. However, it is not clear so far how to express such restrictions within the logic of CQQL.

When being used as a model class in a machine learning context, the restriction to monotone dependencies imposes an inductive bias. If the assumption of monotonicity is valid, this bias is expected to be useful; otherwise, it may prevent one from fitting a proper model. Thus, the choice between the Choquet integral and the weighted DNF as a model class should depend on the application at hand, and on what assumptions can reasonably be made about this application.

For future work, we plan to extend our comparison by including other classes of utility functions, notably the so-called *generalized additive utility models* (Gonzales and Perny 2004). Besides, a generalization of the Choquet integral based on so-called copulas was recently proposed by Kolesarova et al. (2011). Against the background of our study, this extension appears to be especially interesting, as it allows for replacing the minimum as a t-norm in the Moebius representation of the Choquet integral by other t-norms (or copulas). Specifically, by choosing the product as a t-norm, the resulting model would come even closer to the weighted DNF model.

References

Baier D, Brusch M (eds) (2009) Conjointanalyse: Methoden – Anwendungen – Praxisbeispiele. Springer, Berlin/Heidelberg

Baier D, Schmitt I (2012) Logic based conjoint analysis using the commuting quantum query language. In: Proceedings of the joint annual conference of the German Association for Pattern Recognition (DAGM) and the German Classification Society (GfKl), Frankfurt am Main/Marburg

Choquet G (1954) Theory of capacities. Annales de l'institut Fourier 5:131–295

Codd EF (1971) A database sublanguage founded on the relational calculus. In: ACM SIGFIDET workshop on data description, access and control, San Diego, pp 35–68

Domshlak C, Hüllermeier E, Kaci S, Prade H (2011) Preferences in AI: an overview. Artif Intell 175(7–8):1037–1052

Fürnkranz J, Hüllermeier E (eds) (2011) Preference learning. Springer, Berlin/Heidelberg

Gonzales C, Perny P (2004) GAI networks for utility elicitation. In: Proceedings of the 9th international conference on principles of knowledge representation and reasoning, Whistler, pp 224–234

Grabisch M (1997) k-order additive discrete fuzzy measures and their representation. Fuzzy Sets Syst 92(2):167–189

Green PE, Rao VR (1971) Conjoint measurement for quantifying judgmental data. J Mark Res VIII:355–363

Hüllermeier E, Tehrani AF (2012) On the VC dimension of the Choquet integral. In: IPMU-2012, 14th international conference on information processing and management of uncertainty in knowledge-based systems, Catania

Klement E, Mesiar R, Pap E (2002) Triangular norms. Kluwer Academic, Dordrecht/Boston/London

Kolesarova A, Stupnanova A, Beganova J (2011) On extension of fuzzy measures to aggregation functions. In: Proceedings of the Eusflat-2011, 7th international conference of the European Society for fuzzy logic and technology, Aix-les-Bains, pp 28–32

Maier D (1983) The theory of relational databases. Computer Science Press, Rockville

Murofushi T, Soneda S (1993) Techniques for reading fuzzy measures (III): interaction index. In: Proceedings of the 9th fuzzy systems symposium, Sapporo, pp 693–696

von Neumann J (1932) Grundlagen der Quantenmechanik. Springer, Berlin/Heidelberg/New York

Rota G (1964) On the foundations of combinatorial theory I. Theory of Möbius functions. Probab Theory Relat Fields 2(4):340–368. doi:10.1007/BF00531932

Schmitt I (2008) QQL: a DB&IR query language. VLDB J 17(1):39–56

Schmitt I, Zellhöfer D (2012) Condition learning from user preferences. In: Sixth international conference on research challenges in information science, Valencia. IEEE, pp 1–11

Shapley LS (1953) A value for n-person games. In: Kuhn H, Tucker A (eds) Contributions to the theory of games, vol II. Annals of mathematical studies 28. Princeton University Press, Princeton, pp 307–317

Sugeno M (1974) Theory of fuzzy integrals and its application. PhD thesis, Tokyo Institute of Technology

Tehrani AF, Cheng W, Dembczynski K, Hüllermeier E (2012) Learning monotone nonlinear models using the Choquet integral. Mach Learn 89(1):183–211

Vapnik V (1998) Statistical learning theory. Wiley, New York

Zellhöfer D, Schmitt I (2010) A preference-based approach for interactive weight learning: learning weights within a logic-based query language. Distrib Parallel Databases 27(1):31–51

A Symmetry Test for One-Mode Three-Way Proximity Data

Atsuho Nakayama, Hiroyuki Tsurumi, and Akinori Okada

Abstract Recently, several major advances in models of asymmetric proximity data analysis have occurred. These models usually do not deal with the relationships among three or more objects, but instead, those between two objects. However, there exist some approaches for analyzing one-mode three-way asymmetric proximity data that represent triadic relationships among three objects. Nonetheless, a method that evaluates the asymmetry of one-mode three-way asymmetric proximity data has not yet been proposed. There is no measure for judging the necessity of a symmetric model, reconstructed method, or asymmetric model analysis. The present study proposes a method that evaluates the asymmetry of one-mode three-way proximity data. In a square contingency table, a symmetry test is studied to check whether the data are symmetric. We propose a method that extends this symmetry test for square contingency tables to one-mode three-way proximity data.

A. Nakayama (✉)
Graduate School of Social Sciences, Tokyo Metropolitan University, 1-1 Minami-Ohsawa,
Hachioji-shi, Tokyo 192-0397, Japan
e-mail: atsuho@tmu.ac.jp

H. Tsurumi
College of Business Administration, Yokohama National University, 79-4 Tokiwadai,
Hodogayaku, Yokohama 240-8501, Japan
e-mail: tsurumi@ynu.ac.jp

A. Okada
Graduate School of Management and Information Sciences, Tama University, 4-1-1 Hijirigaoka,
Tama City, Tokyo 206-0022, Japan
e-mail: okada@rikkyo.ac.jp; okada@tama.ac.jp

W. Gaul et al. (eds.), *German-Japanese Interchange of Data Analysis Results*,
Studies in Classification, Data Analysis, and Knowledge Organization,
DOI 10.1007/978-3-319-01264-3_11,
© Springer International Publishing Switzerland 2014

1 Introduction

The interest in studies of asymmetric proximity data models has increased, and many significant contributions have been made in this area (e.g. Chino 2012). Asymmetric relationships among objects are common phenomena in marketing research, consumer studies, and other fields. Most previous studies have proposed analyzing one-mode two-way asymmetric proximity data, which show the relationships between two objects. Proximity data may be classified according to the number of directions and modes used (Carroll and Arabie 1980). One-mode two-way asymmetric proximity data describes, for example, the probability of a consumer's switching to brand j, given that brand i was bought on the last purchase. The consumer-switching probability matrix is assumed to have I rows and I columns, where I indexes the same ordered set of I objects for both rows and columns. Carroll and Arabie (1980) labeled such a single matrix two-way because it has both rows and columns. However, both directions correspond to the same set of objects, so it is said that only one mode is included. By contrast, a proximities matrix of I objects by S attributes is considered two-mode two-way data. The two modes are the objects and the source. Then, proximities matrices for objects i and j according to the k-th source are considered two-mode three-way data. One-mode three-way proximity data consist of numerical values assigned to triples of objects.

However, the representations obtained from the analyses of one-mode two-way asymmetric proximity data cannot explain higher-order phenomena with interaction effects caused by more than two influences. Therefore, there is a need for a method that is capable of analyzing asymmetric relationships among three or more objects. Some approaches for analyzing one-mode three-way asymmetric proximity data exist. The easiest is to average the elements in such a one-mode three-way asymmetric proximity data δ_{ijk} as

$$\delta'_{ijk} = (\delta_{ijk} + \delta_{ikj} + \delta_{jik} + \delta_{jki} + \delta_{kij} + \delta_{kji})/6, \tag{1}$$

where i, j, $k = 1, \ldots, n$ denote the objects. One-mode three-way asymmetric proximity data δ_{ijk} are symmetrized into symmetric proximity data δ'_{ijk}. Asymmetric information is lost in this approach. Analyses based on this approach are not able to represent the asymmetric information in one-mode three-way asymmetric proximity data.

In other approaches, such as that of De Rooij and Heiser (2000), asymmetric proximities were directly analyzed by asymmetric models. This approach can represent asymmetric information. De Rooij and Heiser (2000) proposed triadic distance models for the analysis of asymmetric three-way proximity data. In their model, triadic distances d_{ijk} are used that are the function of the dyadic distances

$$d^2_{ijk}(\mathbf{X}; \mathbf{u}; \mathbf{v}) = d^2_{ij}(\mathbf{X}; \mathbf{u}) + d^2_{jk}(\mathbf{X}; \mathbf{v}) + d^2_{ik}(\mathbf{X}; \mathbf{u} + \mathbf{v}), \tag{2}$$

where i, j, $k = 1, \ldots, n$ denote the objects, and d_{ij} is the dyadic distance between objects i and j. The dyadic distance is defined as

$$d_{ij}^2(\mathbf{X}; \mathbf{u}) = \sum_t (x_{it} - x_{jt} + u_t),$$ (3)

where i, $j = 1, \ldots, n$ denote the objects, the matrix $\mathbf{X}$ contains the coordinates x_{it}, and t denotes the dimension ($t = 1, \ldots, T$). Here, i is defined as corresponding to the first way, j to the second way, and k to the third way of one-mode three-way proximity data. De Rooij and Heiser (2000) assumed asymmetry between the first and second ways and asymmetry between the second and third ways. These two asymmetries do not need to be the same. The asymmetry between the first and third way is equal to the sum of the two asymmetries. The asymmetry is modeled by a shift in the Euclidean distance between the two ways. The asymmetry between the first and second way is modeled by a shift $\mathbf{u}$, the asymmetry between the second and third way is modeled by a shift $\mathbf{v}$, and the asymmetry between the first and third way is modeled by a shift $\mathbf{u} + \mathbf{v}$.

In other cases, asymmetric proximities were reconstructed before analysis. Nakayama and Okada (2012) proposed a method to reconstruct one-mode three-way asymmetric data such that the overall sum of the rows, columns, and depths is equal over all objects. Hence, the identity of the marginals is satisfied by reconstruction in the one-mode three-way proximity data. The each way of one-mode three-way asymmetric data is labeled sequentially starting with the rows, followed by columns, and finally depth. Nakayama and Okada (2012)'s method extended that of Harshman et al. (1982) to one-mode three-way asymmetric proximity data. Harshman et al. (1982)'s method made the overall sum of the rows and columns equal over all objects. Nakayama and Okada (2012)'s method added a depth condition. Harshman et al. (1982)'s method is effective for analyzing data that have differences among the overall sum of the rows and columns, depending on external factors. Thus, it is possible to show the structural factors of interest clearly. Nakayama and Okada (2012) proposed a method that extends Harshman et al. (1982)'s method to one-mode three-way asymmetric proximity data. The method reconstructs the one-mode three-way asymmetric proximity data δ_{ijk} as

$$\overline{\delta_{i..}} = \frac{1}{6n} \sum \sum (\delta_{ijk} + \delta_{ikj} + \delta_{jik} + \delta_{jki} + \delta_{kij} + \delta_{kji}) = m$$ (4)

for each i. The value of m is usually set at the grand mean of the unadjusted data. The difference between the sum of each row, column, and depth, which multiplies row i, column i, and depth i by constant c_i, and the grand mean of the unadjusted one-mode three-way asymmetric proximity data may be minimized iteratively to find the constant c_i, which equalizes the sum of each row, column, and depth. The method iteratively calculates constant c_i by the quasi-Newton method under the following conditions:

$$\sum_i (c_i c_1 c_1 \delta_{i11} + \cdots + c_i c_1 c_n \delta_{i1n} + c_1 c_i c_1 \delta_{1i1} + \cdots$$

$$+ c_n c_i c_1 \delta_{ni1} + c_1 c_1 c_i \delta_{11i} + \cdots + c_1 c_n c_i \delta_{1ni})$$

$$= \sum_i (c_i c_2 c_1 \delta_{i21} + \cdots + c_i c_2 c_n \delta_{i2n} + c_1 c_i c_2 \delta_{1i2} + \cdots$$

$$+ c_n c_i c_2 \delta_{ni2} + c_2 c_1 c_i \delta_{21i} + \cdots + c_2 c_n c_i \delta_{2ni})$$

$$= \sum_i (c_i c_n c_1 \delta_{in1} + \cdots + c_i c_n c_n \delta_{inn} + c_1 c_i c_n \delta_{1in} + \cdots$$

$$+ c_n c_i c_n \delta_{nin} + c_n c_1 c_i \delta_{n1i} + \cdots + c_n c_n c_i \delta_{nni})$$

$$= \frac{1}{n} \sum_i \sum_j (\delta_{ij1} + \cdots + \delta_{ijn} + \delta_{1ij} + \cdots + \delta_{nij} + \delta_{j1i} + \cdots + \delta_{jni}). \qquad (5)$$

It is important to select an appropriate approach based on the characteristics of the one-mode three-way asymmetric proximity data. If it cannot be assumed that asymmetry has arisen by chance and it is thought that asymmetry exists, the symmetrized data should not be analyzed. The one-mode three-way asymmetric proximity data should be reconstructed so that the overall sum of the rows, columns, and depths is made equal over all objects if it is thought that the asymmetry and the identity of the marginals exists. Because the identity of the marginals is satisfied by the reconstruction in the one-mode three-way proximity data, the one-mode three-way asymmetric proximity data should not be reconstructed when it is thought that the asymmetry exists but the identity of the marginals does not exist. It is desirable to analyze the asymmetric proximity data by asymmetric models such as that of De Rooij and Heiser (2000). However, a method that evaluates the asymmetry of one-mode three-way asymmetric proximity data has not been proposed.

Here, we propose a method that evaluates the asymmetry of the one-mode three-way proximity data. In a square contingency table, the test of symmetry had previously been studied to check whether the data were symmetric by the chi-square test of symmetry, chi-square test of identity of the marginals, and chi-square test of quasi-symmetry. We propose a method that extends the test of symmetry for square contingency tables to three-way contingency tables. We consider one-mode two-way proximity data as a square contingency table and one-mode three-way proximity data as a three-way contingency table. Then, we examine the necessity for the analysis of the symmetric model, reconstructed method, and asymmetric model by the test of symmetry. We study the chi-square test of symmetry and identity of the marginal in particular detail.

2   The Framework of the Proposed Test of Symmetry

We check the asymmetry of one-mode three-way proximity data by a chi-square test of symmetry. The symmetrized asymmetric data are analyzed by a symmetric model if the result of the chi-square test of symmetry shows that the data are symmetric.

However, asymmetric data should not be symmetrized if the result of the chi-square test of symmetry shows that the data are asymmetric. We examine the identity of the marginals of the one-mode three-way proximity data by a chi-square test of identity of the marginals when the chi-square test of symmetry shows that the symmetry is not satisfied in the one-mode three-way proximity data. It is appropriate to reconstruct asymmetric data by removing the extraneous size differences if the identity of the marginals is satisfied in one-mode three-way proximity data that do not satisfy the symmetry. However, asymmetric data should be analyzed by an asymmetric model such as that of De Rooij and Heiser (2000) if the identity of the marginals is not satisfied.

2.1 Chi-Square Test of Symmetry for One-Mode Three-Way Asymmetric Proximity Data

When two variables have the same number of modalities in a square contingency table, it can be ascertained whether the underlying joint probability distribution is symmetric. In other words, we want to find out whether the probability of the outcome δ_{ij} is the same as the probability of the outcome δ_{ji}. This can be examined by a chi-square test of symmetry for a square contingency table, which assesses the plausibility of the null hypothesis $\mathbf{H_0}: \delta_{ij} = \delta_{ji}$, where δ_{ij} is the similarity between objects i and j. The test requires test statistics whose value for the sample may be considered a fair indicator of the plausibility of the null assumption. The chi-squared statistic for this test is

$$\chi^2 = \sum_i \sum_j \left(\delta_{ij} - \frac{\delta_{ij} + \delta_{ji}}{2} \right)^2 \bigg/ \frac{\delta_{ij} + \delta_{ji}}{2}, \qquad (6)$$

which has $n(n-1)/2$ degrees of freedom for n objects under the null hypothesis of symmetry of the expected counts. Refer to Bowker (1948). The number of the degrees of freedom corresponds to the number of null hypotheses.

We propose a method that extends the chi-square test of symmetry to one-mode three-way asymmetric proximity data. The test assesses the plausibility of the null hypothesis $\mathbf{H_0}: \delta_{ijk} = \delta_{ikj} = \delta_{jik} = \delta_{jki} = \delta_{kij} = \delta_{kji}$, The chi-square statistic is

$$\chi^2 = \sum_i \sum_j \sum_k \left(\delta_{ijk} - \frac{\delta_{ijk} + \delta_{ikj} + \delta_{jik} + \delta_{jki} + \delta_{kij} + \delta_{kji}}{6} \right)^2 \bigg/$$
$$\frac{\delta_{ijk} + \delta_{ikj} + \delta_{jik} + \delta_{jki} + \delta_{kij} + \delta_{kji}}{6}, \qquad (7)$$

which has $n(n-1)(n-2)/6$ degrees of freedom for n under the null hypothesis of symmetry of the expected counts. The number of degrees of freedom corresponds to the number of null hypotheses.

2.2 Chi-Square Test of Marginal Identity in One-Mode Three-Way Asymmetric Proximity Data

In the case of independence, the fundamental parameter behind a square contingency table is the joint probability distribution. The joint probability distribution can be derived from the marginal probability distributions of $\delta_{i.}$ and $\delta_{.j}$. It is therefore natural to inquire whether the two variables have identical distributions. Hence, it is necessary to examine the identity of the marginals by a chi-square test of identity. The identity of the marginals means that average transitions between objects are equal. A chi-square test of identity of the marginals in a square contingency table assesses the plausibility of the null hypothesis $\mathbf{H_0}$: $\delta_{i.} = \delta_{.j}$. The test requires test statistics whose value for a sample may be considered a fair indicator of the plausibility of the null hypothesis. The statistic for this test is

$$\chi^2 = \sum_i \sum_j \left(\delta_{i.} - \frac{\delta_{i.} + \delta_{.j}}{2} \right)^2 \bigg/ \frac{\delta_{i.} + \delta_{.j}}{2}, \tag{8}$$

which has $n - 1$ degrees of freedom for n objects.

We propose a method that extends the chi-square test of marginal identity for one-mode three-way proximity data. The chi-square test of identity of the marginals in one-mode three-way asymmetric proximity data assesses the plausibility of the null hypothesis $\mathbf{H_0}$: $\delta_{ij.} = \delta_{.jk} = \delta_{i.k}$. The chi-square statistic is

$$\chi^2 = \sum_i \sum_j \sum_k \left(\delta_{ij.} - \frac{\delta_{ij.} + \delta_{.jk} + \delta_{i.k}}{3} \right)^2 \bigg/ \frac{\delta_{ij.} + \delta_{.jk} + \delta_{i.k}}{3}, \tag{9}$$

which has $n - 1$ degrees of freedom for n objects.

3 An Application

We applied the proposed method to consecutive Swedish election data obtained from Upton (1978). Swedish respondents were asked how they voted in three consecutive elections (1964, 1968, and 1970). There are four political parties: the Social Democrats (SD), the Center Party (C), the People's Party (P), and the Conservatives (Con). This ordering is from left- to right-wing parties. The data are given in Table 1. The data give the frequency of 64 possible sequences among these four parties at the three time points. The frequency matrices are assumed to have four rows, columns, and depths, where four indexes the same ordered set of

Table 1 Transition frequency data among four parties in three consecutive Swedish elections (1964, 1968, and 1970)

1964	1968	1970 SD	C	P	Con
Social Democrats (SD)	SD	812	27	16	5
	C	5	20	6	0
	P	2	3	4	0
	Con	3	3	4	2
The Center Party (C)	SD	21	6	1	0
	C	3	216	6	2
	P	0	3	7	0
	Con	0	9	0	4
The People's Party (P)	SD	15	2	8	0
	C	1	37	8	0
	P	1	17	157	4
	Con	0	2	12	6
The Conservatives (Con)	SD	2	0	0	1
	C	0	13	1	4
	P	0	3	17	1
	Con	0	12	11	126

four objects for the rows, columns, and depths. Thus, the frequency matrices are considered to be $4 \times 4 \times 4$ one-mode three-way asymmetric proximity data.

For the one-mode three-way asymmetric proximity data, the chi-square test of symmetry assessing the plausibility of the null hypothesis $\mathbf{H_0}$: $\delta_{ijk} = \delta_{ikj} = \delta_{jik} = \delta_{jki} = \delta_{kij} = \delta_{kji}$ was performed. The chi-square statistics were calculated from the transition frequency data. The calculated chi-square statistics were greater than the critical value at a significance level of $\alpha = 0.01$ with four degrees of freedom. Hence, we rejected our null hypothesis $\mathbf{H_0}$ and concluded that the probabilities of the outcomes δ_{ijk}, δ_{ikj}, δ_{jik}, δ_{jki}, δ_{kij}, and δ_{kji} were not the same.

For the transition-frequency data, the chi-square test of identity of the marginals assessing the plausibility of the null hypothesis $\mathbf{H_0}$: $\delta_{ij.} = \delta_{.jk} = \delta_{i.k}$ was performed. The chi-square statistics were calculated from the transition-frequency data. The calculated chi-square statistics were greater than the critical value at a significance level of $\alpha = 0.01$ with three degrees of freedom. Thus, we rejected our null hypothesis $\mathbf{H_0}$ and concluded that the probabilities of the outcomes $\delta_{ij.}$, $\delta_{.jk}$, and $\delta_{i.k}$ were not the same.

From the results of the chi-square test of symmetry, the data were not symmetric. Hence, the transition-frequency data did not need to be symmetrized. However, the results of the chi-square marginal identity test showed that identity of the marginals did not exist. Analysis of the asymmetric model was hence preferable to the analysis of the symmetric model or the analysis of the reconstruction of the one-mode three-way asymmetric data so that the overall sum of the rows, columns, and depths was made equal over all of the objects.

4 Discussion and Conclusion

To check the results of the chi-square tests, we compared the results of the analysis using symmetric, reconstructed, and asymmetric models. First, we analyzed the one-mode three-way asymmetric proximity data that were symmetrized by Eq. (1). Next, we analyzed the one-mode three-way asymmetric proximity data that were reconstructed using the method of Nakayama and Okada (2012). The reconstructed asymmetric proximity data were also symmetrized using Eq. (1). These symmetrized proximity data were analyzed by a generalized Euclidean distance model (e.g. De Rooij and Gower 2003). In a generalized Euclidean distance model, the triadic distances d_{ijk} are defined as

$$d_{ijk} = (d_{ij}^2 + d_{jk}^2 + d_{ik}^2)^{1/2}, \tag{10}$$

where d_{ij} is the Euclidean distance between points i and j representing objects i and j. These analyses used the maximum dimensionalities of categories nine through five. Therefore, the first stress values were obtained in nine- through unidimensional spaces. Similarly, the second stress values were obtained in eight- through unidimensional spaces, the third stress values in seven- through unidimensional spaces, and so on. The smallest stress value in each dimensional space was chosen as the minimum stress value in that dimensional space. The two-dimensional configuration in the present analysis is now discussed. The two-dimensional configuration allows easy understanding of the relationships among the objects. The stress value in two-dimensional space obtained from the reconstructed data was 0.226, and the stress value in two-dimensional space obtained from the non-reconstructed data was 0.300.

Figure 1 shows the two-dimensional configuration obtained from the analysis of reconstructed transition-frequency data by the method of Nakayama and Okada (2012). Figure 2 shows the two-dimensional configuration obtained from the analysis of symmetrized transition-frequency data without reconstructing the data. In Fig. 1, the positions of the parties are based on their characteristics. Left- to right-wing parties are located in clockwise order. The Center party is located near the People's party. The relationships among the parties seem to be correct; it cannot be expected that as many people switch from a right-wing party to a left-wing party as switch from a right-wing or left-wing party to a party in the middle of the political spectrum. However, in Fig. 2, the positions of the parties are not based on their characteristics. Left- to right-wing parties are not located in clockwise order. The Center party is located far from the People's party. The relationships among the parties are approximate.

Next, the two-dimensional results of De Rooij and Heiser (2000)'s one-mode three-way asymmetric model were compared with those of Nakayama and Okada (2012)'s method. De Rooij and Heiser (2000) analyzed the same transition-frequency data of Swedish respondents. The results of De Rooij and Heiser (2000)'s one-mode three-way asymmetric model have the same tendency as those

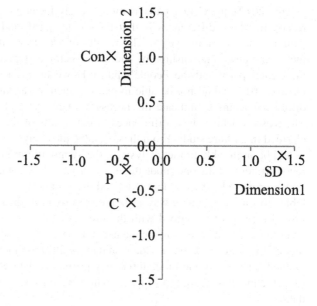

Fig. 1 Two-dimensional configuration obtained from the analysis of reconstructed transition frequency data by Nakayama and Okada (2012)'s method

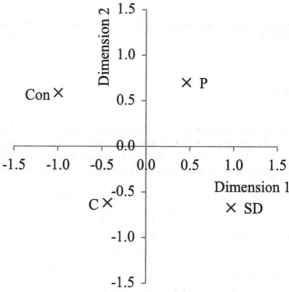

Fig. 2 Two-dimensional configuration obtained from the analysis of symmetrized transition-frequency data without reconstructing the data

of Nakayama and Okada (2012)'s method. The positions of the objects using the De Rooij and Heiser (2000) method are similar to those using the Nakayama and Okada (2012) method. Left- to right-wing parties are located in clockwise order. However, the results of De Rooij and Heiser (2000) partially differ in tendency from those of the Nakayama and Okada (2012) reconstructed method. These differences exist in the Center party and the People's party. In the results of Nakayama and

Okada (2012), the Center party is more closely located to the People's party. The reason for these differences is that the Center party and the People's party are gaining votes over the years. The results of De Rooij and Heiser (2000) show the asymmetric relationships by the slide-vector that pointed in the direction of the Center party and the People's party. However, the results of Nakayama and Okada (2012) would not be able to express such an asymmetric relationship. The difference seems to indicate the necessity of analysis of the asymmetric model. The reasons behind these differences in the results of the two analyses should be clarified in a future study. From these results, the analysis of the asymmetric model would be preferable to the analysis of the symmetric model or the analysis of the reconstruction of the one-mode three-way asymmetric data.

We proposed a method that extended the symmetry test for square contingency tables to one-mode three-way asymmetry proximity data. The proposed method provided results that agreed with the present analysis. The proposed method can be used successfully to evaluate the necessity of a one-mode three-way asymmetric model. In future work, we are interested in establishing the validity of the proposed method for various one-mode three-way proximity datasets. Additionally, we aim to tackle the chi-square test of quasi-symmetry for one-mode three-way proximity data.

Acknowledgements We would like to express our gratitude to two anonymous referees for their valuable reviews.

References

Bowker AH (1948) A test for symmetry in contingency tables. J Am Stat Assoc 43:572–574

Carroll JD, Arabie P (1980) Multidimensional scaling. In: Rosenzweig MR, Porter LW (eds) Annual review of psychology, vol 31. Annual Reviews, Palo Alto, pp 607–649

Chino N (2012) A brief survey of asymmetric mds and some open problems. Behaviormetrika 39:127–165

De Rooij M, Gower JC (2003) The geometry of triadic distances. J Classif 20:181–220

De Rooij M, Heiser WJ (2000) Triadic distances models for the analysis of asymmetric three-way proximity data. Br J Math Stat Psychol 53:99–119

Harshman RA, Green PE, Wind Y, Lundy ME (1982) A model for the analysis of asymmetric data in marketing research. Mark Sci 1:205–242

Nakayama A, Okada A (2012) Reconstructing one-mode three-way asymmetric data for multidimensional scaling. In: Challenges at the interface of data analysis, computer science, and optimization, Karlsruhe. Springer, Berlin/Heidelberg, pp 133–141

Upton GJG (1978) The analysis of cross-tabulated data. Wiley, Chichester

Analysis of Conditional and Marginal Association in One-Mode Three-Way Proximity Data

Atsuho Nakayama

Abstract The purpose of this study was to examine the necessity for one-mode three-way multidimensional scaling analysis. In many cases, the results of the analysis of one-mode three-way multidimensional scaling are similar to those of one-mode two-way multidimensional scaling for lower dimensions, and, in fact, multidimensional scaling can be used for low dimensional analysis. Our results demonstrated that at lower dimensionality, triadic relationships represented by the results of one-mode three-way multidimensional scaling were almost consistent with the dyadic relationships derived from one-mode two-way multidimensional scaling. However, triadic relationships differ from dyadic relationships in analyses of higher dimensionality. The degree of coincidence obtained for one-mode three- and two-way multidimensional scaling revealed that triadic relationships can only be represented by one-mode three-way multidimensional scaling; specifically, triadic relationships based on conditional associations must be separately explained in terms of marginal associations for higher dimensionality analysis.

1 Introduction

Multidimensional scaling (MDS) can be classified according to the number of directions and modes used (Carroll and Arabie 1980). A single-symmetric proximity matrix has I rows and I columns, where I indexes the same ordered set of I objects for both rows and columns. Carroll and Arabie (1980) referred to this kind of single matrix as two-way, because it has both rows and columns; this corresponds to an MDS involving only one inputted matrix as a two-way analysis. Because both directions correspond to the same set of objects, the model only includes one mode.

A. Nakayama (✉)
Tokyo Metropolitan University, 1-1 Minami-Ohsawa, Hachioji-shi, 192-0397, Japan
e-mail: atsuho@tmu.ac.jp

W. Gaul et al. (eds.), *German-Japanese Interchange of Data Analysis Results*,
Studies in Classification, Data Analysis, and Knowledge Organization,
DOI 10.1007/978-3-319-01264-3_12,
© Springer International Publishing Switzerland 2014

However, the two modes have two different directions, such as objects and sources. Therefore, proximity matrices for objects i and i, according to the k-th source, are considered two-mode three-way matrices. One-mode three-way proximity data consist of numerical values assigned to triples of objects.

Researchers have often used one-mode two-way MDS (Kruskal 1964a,b). One configuration, $\mathbf{X}$ of n points $x_i = (x_{i1}, \ldots, x_{ip})$, is assumed, for $i = 1, \ldots, n$, in a p-dimensional Euclidean space, where the x_i-coordinate corresponds to the point for object i. Dyadic distances d_{ij} between two points, representing objects i and j in the configuration, are given by

$$d_{ij} = \left(\sum_{t=1}^{p} \left(x_{it} - x_{jt} \right)^2 \right)^{1/2}. \tag{1}$$

The dyadic distances d_{ij} are determined by finding the $\hat{d}_{ij}$ that satisfies the following conditions:

$$\delta_{ij} < \delta_{rs} \Rightarrow \hat{d}_{ij} > \hat{d}_{rs} \text{ for all } i < j, \ r < s \tag{2}$$

where δ_{ij} represents the one-mode two-way proximity data. The measure of the badness-of-fit of d_{ij} to δ_{ij} is called the stress S and is based on the stress formula defined below (Kruskal and Carroll 1969):

$$S = \sqrt{ \sum_{i<j}^{n} (d_{ij} - \hat{d}_{ij})^2 \Big/ \sum_{i<j}^{n} (d_{ij} - \bar{d}_{ij})^2 }. \tag{3}$$

Using the method of steepest descent, S may be minimized iteratively to find a configuration of minimum stress and the gradient of S, which is given by $\partial S / \partial x_{is}$. Once the gradient has been calculated, it is possible to consider whether a local minimum has been reached. If a local minimum has been reached, then a suitable output is created. However, if a local minimum has not been reached, a new step-size is calculated. From this, a new configuration is calculated, and the iterations begin again.

However, a model capable of analyzing proximities data that differ from one-mode two-way proximities data is needed. A new model is required to explain high-level phenomena among three objects. Previous studies of one-mode three-way MDS usually assumed some functional relationship between one-mode three-way proximity data and the triadic distances (e.g. Gower and De Rooij 2003; De Rooij and Gower 2003; De Rooij 2008), and these functions generally used a linear combination of squared distances. De Rooij and Gower (2003) used symmetric functions of triadic distances, including the perimeter distance, generalized Euclidean distance, generalized dominance distance, variance function, area of the triangle, and the product model. The triadic distances between the points i, j, and k are d_{ijk}. For example, the generalized Euclidean distance model is given by

$$d_{ijk} = (d_{ij}^2 + d_{jk}^2 + d_{ik}^2)^{1/2}. \tag{4}$$

The triadic distances d_{ijk} are determined by finding a $\hat{d}_{ijk}$ that satisfies the following conditions:

$$\delta_{ijk} < \delta_{rst} \Rightarrow \hat{d}_{ijk} > \hat{d}_{rst} \text{ for all } i < j < k, \, r < s < t, \tag{5}$$

where δ_{ijk} represents the one-mode three-way proximity data. The measure of badness-of-fit of d_{ijk} to δ_{ijk} is called the stress S and is based on the stress formula defined below (Kruskal and Carroll 1969)

$$S = \sqrt{\sum_{i<j<k}^{n} (d_{ijk} - \hat{d}_{ijk})^2 \Big/ \sum_{i<j<k}^{n} (d_{ijk} - \bar{d}_{ijk})^2}. \tag{6}$$

Using the method of steepest descent, S may be minimized iteratively to find a configuration of minimum stress and the gradient of S, which is given by $\partial S / \partial x_{is}$.

Gower and De Rooij (2003) reported that the results of a one-mode three-way MDS are likely to resemble those of a one-mode two-way MDS. Here, we propose a method that judges the necessity for analysis of one-mode three-way MDS. We calculated the degree of coincidence for each configuration of one-mode three- and two-way MDS to examine whether the results of a one-mode three-way MDS were strongly associated with the results of a one-mode three-way MDS.

2 The Method

One-mode three-way proximity data consist of numerical values assigned to triples of objects and show triadic relationships. An example of a triadic relationship is the frequency with which goods from each of three categories i, j, and k were purchased simultaneously, or the frequency with which each of the three brands i, j, and k were simultaneously chosen, due to their similarity.

We would usually analyze one-mode three-way proximity data that show the frequencies of co-occurrences between three objects i, j, and k (Table 1a) when one-mode three-way proximity data are calculated from *objects × sources* binary $(0, 1)$ data that displayed co-occurrences among objects, where the value 1 means presence of objects and the value 0 their absence. In this case, the calculation represents the conditional association between objects of first and second way, given objects of third way. However, there is another way to examine one-mode three-way proximity data. One-mode three-way proximity data can be used to show the frequencies of co-occurrences between objects of first and second way, without objects of third way, and can also express the conditional association (Table 1b). When these two one-mode three-way data are combined, we get marginal one-mode

Table 1 Three types of one-mode three-way proximity data. The left and center tables list conditional one-mode three-way proximity data. The left table (a) lists the frequencies of co-occurrences among objects of first, second, and third way. The center table (b) lists the frequencies of co-occurrences between objects of first and second way, without occurrence of object of third way. The right table (c) lists marginal one-mode three-way proximity data

(a)

Third way	First way	Second way i	j	k
Object i	i	241	51	34
	j	51	51	2
	k	34	2	34
Object j	i	51	51	2
	j	51	316	123
	k	2	123	123
Object k	i	34	2	34
	j	2	123	123
	k	34	123	456

(b)

Third way	First way	Second way i	j	k
Object i	i	0	0	0
	j	0	1,265	121
	k	0	121	422
Object j	i	190	0	32
	j	0	0	0
	k	32	0	333
Object k	i	207	49	0
	j	49	1,193	0
	k	0	0	0

(c)

Third way	First way	Second way i	j	k
Object i	i	241	51	34
	j	51	1,316	122
	k	34	123	456
Object j	i	241	51	34
	j	51	1,316	123
	k	34	123	456
Object k	i	241	51	34
	j	51	1,316	123
	k	34	123	456

Table 2 One-mode two-way proximity data

	Second way		
First way	Object i	Object j	Object k
Object i	241	51	34
Object j	51	1,316	122
Object k	34	123	456

three-way proximity data (Table 1c). Marginal one-mode three-way proximity data show the frequencies of co-occurrences among two objects, which do not depend on object of third way. Marginal one-mode three-way proximity data correspond to one-mode two-way proximity data placed as many times as the number of objects. When marginal one-mode three-way proximity data are collapsed, we obtain one-mode two-way proximity data (Table 2). We can treat one-mode two-way proximity data as marginal associations of dyadic relationships in one-mode three-way proximity data.

Gower and De Rooij (2003) reported that the results of a one-mode three-way MDS are likely to resemble those of a one-mode two-way MDS. If the results of a one-mode three-way MDS are almost consistent with those of one-mode two-way MDS, the triadic relationships in one-mode three-way proximity data are almost consistent with the dyadic relationships in one-mode two-way proximity data. The triadic distances obtained from the analysis of one-mode three-way proximity data are consistent with the triadic distances calculated from the coordinates obtained from the analysis of one-mode two-way proximity data. Therefore, the coordinates X_t of n points in a p-dimensional Euclidean space, obtained from the analysis of one-mode three-way proximity data, are almost consistent with the coordinates X_d obtained from the analysis of one-mode two-way proximity data when Procrustes analysis (e.g. Sibson 1978) was used to match the configuration from the one-mode two-way MDS to that for one-mode three-way MDS. We take as our measure of consistency the degree of coincidence of corresponding columns of X_t and X_d .

$$\phi = \sum_{p=1}^{n} X_{tp} X_{dp} \bigg/ \left(\sum_{p=1}^{n} X_{tp} \sum_{p=1}^{n} X_{dp} \right)^{1/2} . \tag{7}$$

Procrustes statistics (e.g. Sibson 1978) are applied to our measure of consistency the degree of coincidence. The degree of coincidence is the sum of products of corresponding rows of X_t and X_d, and ranges from 0 to 1 due to normalization by the sum of the squares. The corresponding columns of X_t and X_d are exactly consistent with each other if the degree of coincidence, ϕ, is 1. The triadic relationships can be also expressed by one-mode two-way MDS. Conditional associations as triadic relationships in one-mode three-way proximity data are considered marginal associations of dyadic relationships in one-mode three-way proximity data. As the degree of coincidence ϕ decreases, the degree of coincidence of the corresponding columns of X_t and X_d also decreases. If the triadic relationships exhibit little consistency with the dyadic relationships, then the results of one-mode three-way MDS are less similar to those of one-mode two-way MDS. The coordinates X_t obtained from the analysis of one-mode three-way proximity data exhibit little consistency with respect to the coordinates X_d obtained from the analysis of one-mode two-way proximity data. The degree of coincidence, ϕ, decreases toward 0. Marginal and conditional associations should be separately considered. Consequently, the triadic relationships can be only represented by one-mode three-way MDS. To determine whether the triadic relationships are strongly associated with the dyadic relationships, we calculated the degree of coincidence, ϕ, between each configuration of one-mode three- and two-way MDS. We examined the necessity for analysis of one-mode three-way MDS based on the degree of coincidence, ϕ.

3   The Analysis

We used binary data that displayed co-occurrences among three beer brands, which were chosen from ten brands, due to their similar brand images. Ten beer brands from four different companies were evaluated based on five attributes: category, taste, malt, price, and history (Table 3). The major producers were Asahi, Kirin, Sapporo, and Suntory, while small local breweries supplied distinct tasting beers. Based on the Japanese taxation system, the varieties of brewed malt beverages in Japan were further categorized into three groups: beer, Happoshu, and third-category beer. Alcoholic beverages based on malt are classified as beer if the weight of the malt extract exceeds 67 % malt. Beer can be further divided into two classes: regular beer and premium beer. The brewed malt beverages market has been declining, while the premium beer market has been expanding, due to the trend toward a more enriched spiritual life. Premium beer is made from rich, pure malt using carefully selected ingredients and original brewing methods. The price of premium beer is higher than that of others, due to the higher quality ingredients. Happoshu is a tax category of Japanese liquor with less than 67 % malt content. This alcoholic beverage is popular among consumers because it is taxed less than beverages legally classified as beer. With alcohol tax revenues decreasing as a result of Happoshu's popularity, the Japanese government eventually raised the nation's tax on low-malt beers. Brewers followed suit by lowering the malt content of their products. Since 2004, Japanese breweries have produced even lower-taxed and non-malt brews made from soybeans and other ingredients, which do not fit the classifications for beer or Happoshu. These lower-taxed and non-malt brews, referred to by the mass media as third-category beer, were developed to compete with Happoshu. The Happoshu market has been declining since the late 2000s, due to the expansion of the third-category beer market.

To obtain background data about the Japanese beer market, a brand image survey of college students was conducted to assess consumers' impressions of various beer brands. College students were asked to select similar brands from a

Table 3 Ten beer brands from four different companies and their characteristics

	Category	Taste	Malt	Price	History
Brand 1 (Kirin)	Regular beer	Mild	Not-all malt	Middle price	Traditional brand
Brand 2 (Kirin)	Regular beer	Rich	All malt	Middle price	New brand
Brand 3 (Asahi)	Regular beer	Mild	Not-all malt	Middle price	New brand
Brand 4 (Sapporo)	Regular beer	Mild	Not-all malt	Middle price	Traditional brand
Brand 5 (Suntory)	Premium beer	Rich	All and pure malt	Premium price	New brands
Brand 6 (Sapporo)	Premium beer	Rich	All and pure malt	Premium price	New brands (resale)
Brand 7 (Kirin)	Third category beer	Mild	Other than malt	Low price	New brand
Brand 8 (Asahi)	Third category beer	Mild	Other than malt	Low price	New brand
Brand 9 (Suntory)	Third category beer	Mild	Other than malt	Low price	New brand
Brand 10 (Sapporo)	Third category beer	Mild	Other than malt	Low price	New brand

list of ten beer brands sold in the Japanese market (Table 3) after watching each beer's TV commercial. From these results, we selected binary data that displayed co-occurrences among three beer brands, chosen from the ten brands due to their similar brand images. One-mode three- and two-way similarity data were calculated from these binary data. The $10 \times 10 \times 10$ one-mode three-way similarity data indicated the frequencies with which each of the three brands was simultaneously chosen, due to their similarity. The 10×10 one-mode two-way similarity data indicated the frequencies with which each of the two brands was simultaneously chosen, due to their similarity. The one-mode three-way proximity data were then analyzed by one-mode three-way MDS, based on a generalized Euclidian distance model (e.g. De Rooij and Gower 2003). The generalized Euclidian distance model was selected because the triadic distances allowed for easy expansion of the one-mode two-way algorithms to that of the model; this provided better visualization of the relationships among objects as a way to better understand them. Next, the one-mode two-way proximity data were analyzed by one-mode two-way MDS (Kruskal 1964a,b). These analyses used the maximum dimensionalities of categories eight through four. Therefore, the first stress values were obtained in eight-through unidimensional spaces. Similarly, the second stress values were obtained in seven- through unidimensional spaces, and so on. The lowest stress value in each

Table 4 Degree of coincidence values, ϕ, for the corresponding dimensions of the configuration of one-mode three- and two-way MDS for each dimensionality

	Dimensionality 5	Dimensionality 4	Dimensionality 3	Dimensionality 2
Dimension 1	0.995	0.993	0.993	0.988
Dimension 2	0.955	0.960	0.954	0.912
Dimension 3	0.827	0.877	0.797	
Dimension 4	0.907	0.973		
Dimension 5	0.666			

dimensional space was chosen as the minimum stress value in that dimensional space. The resulting minimum stress values obtained from the analysis of one-mode three-way proximity data in five- through unidimensional spaces were 0.303, 0.303, 0.325, 0.354, and 0.467, respectively. The resulting minimum stress values obtained from the analysis of one-mode two-way proximity data in five- through unidimensional spaces were 0.000, 0.000, 0.031, 0.144, and 0.280, respectively.

Procrustes analysis was used to match the each dimensional configuration from the one-mode two-way MDS to that for one-mode three-way MDS. The degree of coincidence, ϕ, of corresponding dimensions of the configuration of one-mode three- and two-way MDS was then calculated. The calculation used the results of dimensionality of categories five through one. Therefore, the first degree of coincidence, ϕ, of the corresponding dimensions of the configuration of one-mode three- and two-way MDS was calculated in five- through unidimensional spaces. Similarly, the second degree of coincidence, ϕ, was calculated in four- through unidimensional spaces, and so on. Table 4 lists the resulting degrees of coincidence, ϕ, for the corresponding dimensions of configurations of one-mode three- and two-way MDS for each dimensionality. We examined the degrees of coincidence, ϕ, of the corresponding dimensions of the configuration of one-mode three- and two-way MDS. Every degree of coincidence of one and two dimensions was greater than 0.9. The corresponding dimensions of each configuration of one-mode three- and two-way MDS were consistent. The number of degrees of coincidence less than 0.9 increased as the dimension increased. Some dimensions of each configuration of one-mode three- and two-way MDS were not consistent.

4 Conclusions and Outlook

The configuration obtained from one-mode three-way MDS almost fit that of one-mode two-way MDS in one and two dimensions. However, the coordinate values of some brands of the one-mode three-way MDS did not correspond closely to those of one-mode two-way MDS at higher dimensions. For lower dimensions, the triadic relationships found using one-mode three-way MDS were almost consistent with the dyadic relationships obtained from the results of the one-mode two-way MDS. However, the triadic relationships differed from the dyadic relationships at

Fig. 1 Two-dimensional configuration obtained from the two-dimensional solution and joint representation of the configuration obtained from one-mode three- and two-way MDS

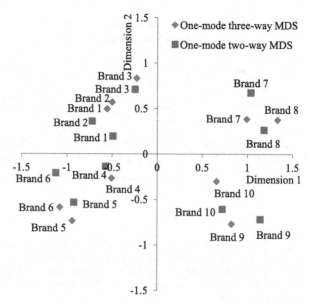

higher dimensions. Triadic relationships based on conditional associations cannot be explained in terms of the dyadic relationships based on marginal associations. They can be represented only by one-mode three-way MDS. The reasons for the inconsistency at higher dimensions were explored by comparing the results of dimensionalities 2 and 3.

Figure 1 presents the two-dimensional configuration of the results of dimensionality 2 and jointly represents the configuration obtained from one-mode three- and two-way MDS. This configuration reveals the similarities among the ten brands. The configuration of one-mode three-way MDS reveals the triadic relationships between each set of three brands that share similar impressions. The configuration of one-mode two-way MDS reveals the dyadic relationships between each set of two brands that share similar impressions. The one-mode three-way MDS configuration had almost the same tendencies as one-mode two-way MDS. Very little distinguished the two configurations for a dimensionality of 2. The brands located in the central portion of Fig. 1 were considered similar to those of the other brands located at the edge of Fig. 1, which were considered dissimilar to the other brands. In Fig. 1, regular beer (brands 1, 2, 3, 4, 5, and 6) are displayed in the left half, and third-category beer (brands 7, 8, 9, and 10) are displayed in the right half of the configuration. The horizontal dimension 1 of the solution is the regular beer versus third-category beer dimension. In Fig. 1, the brands of companies with a smaller market share (brands 4, 5, 6, 9, and 10) are displayed in the lower half, and brands of companies with a larger market share (brands 1, 2, 3, 7, and 8) are displayed in the upper half. Asahi holds the largest share of the present Japanese beer market; Kirin previously held the largest share. The vertical dimensionality two represents

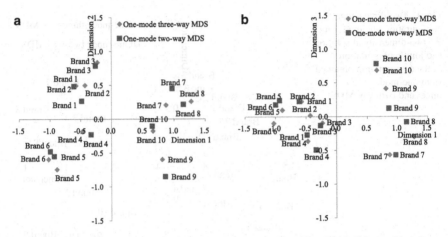

Fig. 2 Two-dimensional view of the three-dimensional configuration. The left figure (**a**) defines the plane by dimensions 1 and 2, and presents the joint representation of the configuration obtained from one-mode three- and two-way MDS. The right figure (**b**) defines the plane by dimensions 1 and 3, and presents the joint representation of the configuration obtained from one-mode three- and two-way MDS

the dimension of brands of companies with a smaller market share versus those with a larger market share.

The three-dimensional configuration of the results of dimensionality three is represented separately. It is divided into configurations of dimensions 1 and 2 (Fig. 2a) and those of dimensions 1 and 3 (Fig. 2b). Figure 2a presents a two-dimensional view of the three-dimensional configuration (which defines the plane by dimensions 1 and 2), as well as the joint representation of the configurations obtained from one-mode three- and two-way MDS. As with the two-dimensional configuration obtained from the two-dimensional solution, little difference was observed between the one-mode three- and two-way MDS configurations. The two configurations displayed almost the same tendencies in terms of similarity among the ten brands. The horizontal dimension 1 of the solution represents regular beer versus third-category beer. The vertical dimension 2 represents the brands of companies with a smaller market share versus those with a larger market share.

Figure 2b presents a two-dimensional view of the three-dimensional configuration, which defines the plane by dimensions 1 and 3, as well as the joint representation of the configuration obtained from one-mode three- and two-way MDS. Some differences appeared in the configuration of one-mode three-way MDS and that of one-mode two-way MDS in Fig. 2b. The configuration of one-mode three-way MDS contains information that cannot be expressed by one-mode two-way MDS. In the vertical dimension 3 of the configuration of one-mode three-way MDS, brands generating an impression of high quality (3, 4, 6, 7, and 8) have a negative value; whereas those generating an impression of good taste (1, 2, 5, 9, and 10) have a positive value. The vertical dimension 3 of the configuration of one-mode three-way MDS represents brands considered to be high quality versus brands considered to be taste good. However, the tendency of vertical dimension 3 for one-mode two-way MDS is less clear than that for one-mode three-way MDS. The coordinate values of brands 1 and 6 reverse the positive and negative values in the vertical dimension 3; see Fig. 2b.

The degree of coincidence and configuration obtained from one-mode three- and two-way MDS demonstrate that triadic relationships can only be represented by one-mode three-way MDS in high dimensionality. In the analysis of lower dimensionality, the triadic relationships that are represented by the results of one-mode three-way MDS are almost consistent with the dyadic relationships derived from the results of one-mode two-way MDS. It is desirable to clarify the insight of low dimensionality based on the results of one-mode two-way MDS. This helps clarify the relationships among objects more easily than the results of one-mode three-way MDS. Additionally, the software required for this approach, such as SAS, R, or SPSS, is more popular than that used for one-mode three-way MDS. However, triadic relationships differ from dyadic relationships in analyses of higher dimensionality. Triadic relationships cannot be partially explained in terms of dyadic relationships. From these results, we can conclude that triadic relationships based on conditional associations must be separately explained in terms of marginal associations in analyses of higher dimensionality. The dataset used in this study required analysis of one-mode three-way MDS. We actively applied our findings about the differences between triadic and dyadic relationships in analyses of higher dimensionality In the future, we plan to establish the validity of the proposed method for various one-mode three-way proximity data. We will also investigate mathematical conditions and assumptions for the proposed method. The reasons for the resemblance between the results of a one-mode three- and two-way MDS are attributed to one-mode three-way MDS, which assumes the functional relationship between triadic relationships and distances is a linear combination of squared distances. If triadic relationships are correlated with dyadic relationships, then the linear functional assumption is not sufficient to explain the triadic relationships. A model representing a nonlinear quadratic form is needed to analyze one-mode three-way proximity data. We hope to address these issues in future studies.

Acknowledgements We would like to express our gratitude to two anonymous referees for their valuable reviews. Some parts of this research were conducted by Nakayama when he was at Nagasaki University.

References

Carroll JD, Arabie P (1980) Multidimensional scaling. In: Annual review of psychology, vol 31. Annual reviews, Palo Alto, pp 607–649

De Rooij M (2008) The analysis of change, newton's law of gravity and association models. J R Stat Soc A (Stat Soc) 171:137–157

De Rooij M, Gower JC (2003) The geometry of triadic distances. J Classif 20(2):181–220

Gower JC, De Rooij M (2003) A comparison of the multidimensional scaling of triadic and dyadic distances. J Classif 20(1):115–136

Kruskal JB (1964a) Multidimensional scaling by optimizing goodness of fit to a nonmetric hypothesis. Psychometrika 29:1–27

Kruskal JB (1964b) Nonmetric multidimensional scaling: a numerical method. Psychometrika 29:115–129

Kruskal JB, Carroll JD (1969) Geometrical models and badness-of-fit functions. In: Multivariate analysis, vol 2. Academic, New York, pp 639–671

Sibson R (1978) Studies in the robustness of multidimensional scaling: procrustes statistics. J R Stat Soc B 40(2):234–238

Analysis of Asymmetric Relationships Among Soft Drink Brands

Akinori Okada

Abstract Brand switching data among eight soft drink brands were analyzed. The data are represented by an 8×8 brand switching matrix. The brand switching matrix is inevitably asymmetric, because the relationship from brand j to brand k is not necessarily equal to the relationship from brand k to brand j. The brand switching matrix was analyzed by asymmetric multidimensional scaling based on singular value decomposition. The four-dimensional result was chosen as the solution. The solution gives the outward tendency, which represents the strength of switching from a corresponding brand to the other brands along each dimension, and the inward tendency, which represents the strength of switching to a corresponding brand from the other brands along each dimension. The solution disclosed that the differences between diet and non-diet brands as well as between cola and lemon-lime brands played important roles in the brand switching.

1 Introduction

Several researchers (Bass et al. 1972; Borg and Groenen 2005, Chap. 23; DeSarbo and Soete 1984; DeSarbo et al. 1990; Zielman and Heiser 1991) have analyzed brand switching data among soft drink brands. The brand switching among n brands is represented by an $n \times n$ brand switching matrix, where the row corresponds to the brand from which the brand switching is made and the column corresponds to the brand to which the brand switching is made. The brand switching matrix is intrinsically asymmetric, because the frequency of the brand switching from brand j to brand k is not necessarily equal to that from brand k to brand j.

A. Okada (✉)
Graduate School of Management and Information Sciences, Tama University, 4-1-1 Hijirigaoka, Tama City, Tokyo 206-0022, Japan
e-mail: okada@rikkyo.ac.jp; okada@tama.ac.jp

W. Gaul et al. (eds.), *German-Japanese Interchange of Data Analysis Results*, Studies in Classification, Data Analysis, and Knowledge Organization, DOI 10.1007/978-3-319-01264-3_13,
© Springer International Publishing Switzerland 2014

Earlier studies focused their attention mainly on how to represent the asymmetric relationships among soft drink brands. The purpose of the present study is not only to represent asymmetric relationships among soft drink brands but also to disclose the competitive relationships among soft drink brands by asymmetric multidimensional scaling.

2 The Data

Brand switching data among eight soft drink brands (Bass et al. 1972), which are represented by an 8×8 brand switching matrix, are analyzed in the present study. The brand switching matrix is analyzed by asymmetric multidimensional scaling (Okada 2008, 2011; Okada and Tsurumi 2012) based on singular value decomposition. The eight soft drink brands are characterized by two attributes; cola/lemon-lime and diet/non diet shown in Table 1.

The data (Bass et al. 1972) have already been normalized by dividing each row element by the sum of the corresponding row elements. Each element represents the proportion of the brand corresponds to the row which is switched to the brand corresponds to the column. And the sum of row elements is unity. The higher the proportion is, the closer the brand corresponding to the row is to the brand corresponding to the column in brand switching. Thus we regard the proportion as the similarity from the brand corresponding to the row to the brand corresponding to the column in brand switching among eight brands.

3 The Method

Brand switching data were analyzed by asymmetric multidimensional scaling (Okada 2008, 2011; Okada and Tsurumi 2012) based on singular value decomposition. Let $\mathbf{A}$ be a matrix of switching or similarity from row brands to column brands. The (j, k) element of $\mathbf{A}$, a_{jk}, represents the switching or similarity from brand j to brand k. The (k, j) element of $\mathbf{A}$, a_{kj}, represents the switching or similarity from brand k to brand j. The two conjugate elements; a_{jk} and a_{kj}, are not always equal. When we have n brands, $\mathbf{A}$ is an $n \times n$ asymmetric matrix. The singular value decomposition of $\mathbf{A}$ is

$$\mathbf{A} = \mathbf{XDY'}, \tag{1}$$

where s$\mathbf{X}$ is the $n \times n$ matrix of left singular vectors of unit length, $\mathbf{D}$ is the $n \times n$ diagonal matrix of singular values in descending order of its diagonal elements, and $\mathbf{Y}$ is the $n \times n$ matrix of right singular vectors of unit length. The j-th element of the i-th column of $\mathbf{X}$, x_{ji}, represents the outward tendency of brand j along dimension i (Okada 2008, 2011). The k-th element of the i-th column of $\mathbf{Y}$, y_{ki}, represents the inward tendency of brand k along dimension i (Okada 2008, 2011).

Table 1 Eight soft drink brands

Brand	Cola/lemon-lime	Diet/non-diet
Coke	Cola	Non-diet
7-Up	Lemon-lime	Non-diet
Tab	Cola	Diet
Like	Lemon-lime	Diet
Pepsi	Cola	Non-diet
Sprite	Lemon-lime	Non-diet
Diet Pepsi	Cola	Diet
Fresca	Lemon-lime	Diet

The outward tendency of a brand represents the strength of switching from the corresponding brand to the other brands in brand switching. The inward tendency of a brand represents the strength of switching to the corresponding brand from the other brands in brand switching. The outward tendency of a brand represents how weak the brand is, and the inward tendency represents how strong the brand is in brand switching.

As stated earlier the (j, k) element of $\mathbf{A}$, a_{jk}, shows the proportion of the brand corresponds to row j which is switched to the brand corresponds to column k, and represents the similarity from object j to object k. The similarity from brand j to brand k, a_{jk}, is approximated by using $m\,(m < n)$ singular vectors and values

$$a_{jk} \cong \sum_{i=1}^{m} d_i x_{ji} y_{ki}, \tag{2}$$

where d_i is the i-th diagonal element of $\mathbf{D}$ or the i-th largest singular value of $\mathbf{A}$. The i-th term of the right side of Eq. (2), $d_i x_{ji} y_{ki}$, is the product of x_{ji}, the outward tendency of brand j, and y_{ki}, the inward tendency of brand k, along dimension i multiplied by the i-th singular value d_i. This term is geometrically represented as the area of a rectangle $(x_{ji} \times y_{ji})$ with positive or negative sign in a plane spanned by the ith left singular vector (abscissa) and the ith right singular vector (ordinate) multiplied by d_i, and represents the ith component of the similarity from brand j to brand k along dimension i, which can be positive or negative (Okada 2011). This means that the algebraic sum of positively or negatively signed areas of rectangles, each comes from each dimension, approximates the similarity a_{jk} (see Fig. 5). The larger m is, the more precisely a_{jk} is approximated. But the larger m needs more terms to approximate a_{jk}, which leads to the complexity. Thus m is determined by balancing the precision of the approximation with the complexity together with the interpretation of the obtained result.

4 Results

Singular value decomposition of the brand switching matrix among eight soft drink brands was computed. Singular values were 1.087, 0.496, 0.406, 0.365, 0.268, 0.197, 0.122, and 0.000. The four-dimensional result was chosen as the solution

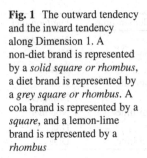

Fig. 1 The outward tendency and the inward tendency along Dimension 1. A non-diet brand is represented by a *solid square or rhombus*, a diet brand is represented by a *grey square or rhombus*. A cola brand is represented by a *square*, and a lemon-lime brand is represented by a *rhombus*

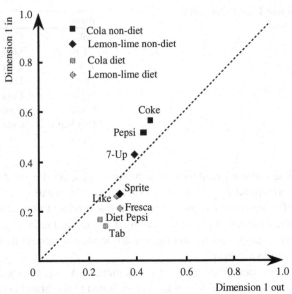

($m = 4$). The proportion of the sum of four squared singular values to the sum of all squared singular values is 0.93. This figure is large enough to represent the asymmetric relationships among eight soft drink brands, and the difference between the fourth and fifth singular value (0.097) is relatively larger than that between the third and fourth singular value (0.041). And the four dimensional result is easy to interpret.

Figure 1 shows the outward and inward tendencies along Dimension 1. The eight soft drink brands are almost along a 45° line emitting from the origin to the upper right direction (which shows the outward tendency is equal to the inward tendency, and is represented by a dotted line in Fig. 1). This tells that the asymmetry in the brand switching along Dimension 1 is not large. The range of the outward tendency is smaller than that of the inward tendency, this comes from the normalization of the row so that the sum of each row elements is unity. The three brands (Coke, 7-Up and Pepsi) have inward tendencies which are larger than their own outward tendencies, suggesting that the three brands are stronger than the other brands in the brand switching along Dimension 1. The three brands are non-diet. Non-diet brands have larger outward and inward tendencies than diet brands have. The eight soft drink brands have positive outward and inward tendencies. The product of the outward tendency of brand j and the inward tendency of brand k is positive. The product multiplied by the first singular value, $d_1 x_{j1} y_{k1}$, represents the similarity from brand j to brand k along Dimension 1. Dimension 1 seems to represent the general dominance or strength of the soft drink brands in the brand switching, because the first singular value is large compared with the other singular values.

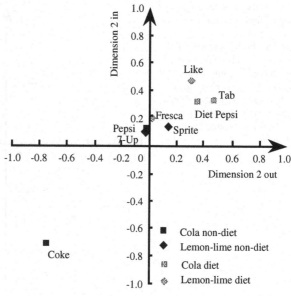

Fig. 2 The outward tendency and the inward tendency along Dimension 2. A non-diet brand is represented by a *solid square or rhombus*, a diet brand is represented by a *grey square or rhombus*. A cola brand is represented by a *square*, and a lemon-lime brand is represented by a *rhombus*

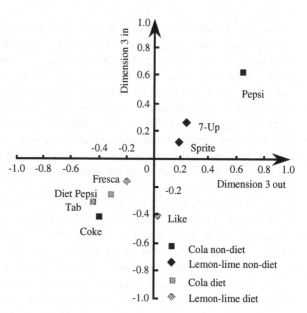

Fig. 3 The outward tendency and the inward tendency along Dimension 3. A non-diet brand is represented by a *solid square or rhombus*, a diet brand is represented by a *grey square or rhombus*. A cola brand is represented by a *square*, and a lemon-lime brand is represented by a *rhombus*

Figures 2–4 show the results along Dimensions 2, 3, and 4, respectively. Contrary with the outward and inward tendencies along Dimension 1, those along Dimensions 2, 3, and 4 are not always positive. Figure 2 shows the result along Dimension 2. Coke is in the third quadrant, while five others are in the first quadrant and two others are in the second quadrant. Dimension 2 differentiates Coke from the other

Fig. 4 The outward tendency
and the inward tendency
along Dimension 4. A
non-diet brand is represented
by a *solid square or rhombus*,
a diet brand is represented by
a *grey square or rhombus*. A
cola brand is represented by a
square, and a lemon-lime
brand is represented by a
rhombus

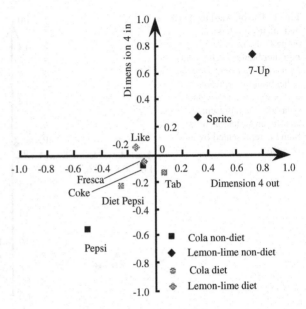

brands, and classifies the brands almost into two groups; one consists of Coke, and the other consists of brands in the first quadrant. The similarity among brands in the first quadrant is positive for any two brands along Dimension 2, because the outward and the inward tendencies are positive. The similarity between a brand in the first quadrant and Coke (in the third quadrant) is negative along Dimension 2, because a brand in the first quadrant and Coke have the outward and the inward tendencies which have opposite signs. The brands in the second quadrant have negative outward tendencies and positive inward tendencies. Dimension 2 outward (the horizontal dimension) seems to differentiate Coke from the other brands. Dimension 2 inward (the vertical dimension) seems to represent the difference between non-diet brands and diet brands.

The results along Dimension 3 is shown in Fig. 3. While Like is in the fourth quadrant, Dimension 3 also classifies the brands almost into two groups; one group consists of brands in the first quadrant, and the other group consists of brands in the third quadrant, where the similarity among brands in these quadrants within the group is positive and the similarity among brands when two brands belong to different groups is negative. Dimension 3 outward (the horizontal dimension) seems to represent three categories; cola brands, lemon-lime brands, and Pepsi. Dimension 3 inward (the vertical dimension) seems to represent other three categories; Coke, diet brands and non-diet brands.

Figure 4 shows the outward and inward tendencies along Dimension 4. While Like is in the second quadrant and Tab is in the fourth quadrant, Dimension 4 also classifies the brands almost into two groups; one group consists of brands in the first quadrant, and the other group consists of brands in the third quadrant, where the similarity among brands within the group is positive and the similarity among brands when two brands belong to different groups is negative. Dimension 4 outward (the horizontal dimension) seems to represent three categories; Pepsi, diet brands and non-diet lemon-lime brands. Dimension 4 inward (the vertical dimension) seems to represent the difference between cola brands and lemon lime brands.

5 Discussion

Brand switching data among eight soft drink brands was analyzed by asymmetric multidimensional scaling using singular value decomposition (Okada 2011; Okada and Tsurumi 2012). Asymmetric multidimensional scaling gives the outward tendency and the inward tendency along each dimension. The outward tendency of a brand represents the strength of switching from the brand to the other brands, and the inward tendency of a brand represents the strength of switching to the brand from the other brands.

In the present model, outward tendency along dimension i, which represents the strength of switching from the brand to the other brands along dimension i (cf. shub weight in Kleinberg (1999)), is represented by the ith left singular vector, and inward tendency along dimension i, which represents the strength of switching to the brand from the other brands along dimension i (cf. authority weight in Kleinberg (1999)), is represented by the ith right singular vector. As shown in Eq. (2), the similarity from brand j to brand k along dimension i is represented by the product of outward tendency of brand j along dimension i x_{ji}, and inward tendency of object j along dimension i y_{ki}, multiplied by the i-th singular value d_i. While the product along Dimension 1 is positive, the products along Dimensions 2, 3, and 4 can be negative. The similarity from brand j to brand k is approximated by the algebraic sum of the products (Eq. 2). The negative terms reduce the similarity. Figure 5 gives an explanation of these.

Figure 5 a–d respectively show the rectangle which corresponds to the similarity along Dimensions 1, 2, 3, and 4. The similarity from Coke to Pepsi along Dimension 1 is represented by the area of the rectangle (the product of the outward tendency of Coke and the inward tendency of Pepsi along Dimension 1) with diagonal stripes from lower left to upper right (Fig. 5a) multiplied by d_1 (1.087). On the other hand, the similarity from Pepsi to Coke along Dimension 1 is represented by the area of the rectangle (the product of the outward tendency of Pepsi and the inward tendency of Coke along Dimension 1) with diagonal stripes from upper left to lower right (Fig. 5a) multiplied by d_1. The similarity from Coke to Pepsi along Dimension 2 is represented by the area of the rectangle (the product of the outward tendency of Coke and the inward tendency of Pepsi along Dimension 2) of diagonal stripes from lower left to upper right multiplied by d_2 (0.496). The product of the outward tendency of Coke and the inward tendency of Pepsi along Dimension 2 is negative, because the outward tendency of Coke is negative and the inward tendency of Pepsi is positive. The area of the rectangle corresponding to the similarity from Coke to

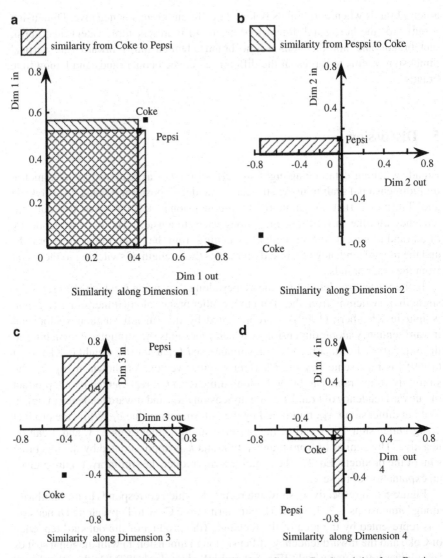

Fig. 5 The rectangle corresponding to the similarity from Coke to Pepsi, and that from Pepsi to Coke along Dimensions 1 (**a**), 2 (**b**), 3 (**c**), and 4 (**d**) respectively

Pepsi is negatively signed. The area of the rectangle corresponding to the similarity from Pepsi to Coke is positively signed (the outward tendency of Pepsi and the inward tendency of Coke are negative). The area of the rectangles corresponding to the similarity from Coke to Pepsi and that from Pepsi to Coke along Dimension 3 are negatively signed. The area of the rectangles corresponding to the similarity from Coke to Pepsi and that from Pepsi to Coke along Dimension 4 are positively signed. The similarity from Coke to Pepsi is represented by the algebraic sum of positively or negatively signed areas formed by the outward tendency of Coke and the inward tendency of Pepsi along Dimensions 1, 2, 3, and 4 multiplied by singular values. In the case of the similarity from Coke to Pepsi, areas of rectangles along Dimensions 2 and 3 are negatively signed, and these two terms reduce the similarity.

Along Dimensions 2, 3, and 4 the outward tendency and the inward tendency can be negative. This means that the product of the outward and inward tendencies is positive when the two have the same sign, and is negative when the two have opposite signs. This tells that (a) the similarity between any two brands in the first and the third quadrants is positive, (b) the similarity between any two brands in the second and the fourth quadrants is negative, (c) the similarity between one brand in the first quadrant and the other in the third quadrant is negative, and (d) the similarity between one brand in the second quadrant and the other in the fourth quadrant is positive. This also tells that the similarity from brand j to brand k is smaller than the similarity from brand k to brand j, when the angle (of two lines connecting the origin and the points representing brand k and brand j, respectively) from brand j to brand k is positive (counter clockwise), suggesting brand k dominates over brand j. These can represent the competitive relationships among brands described below.

Along Dimension 1, the eight soft drink brands are almost along a 45° line emitting from the origin to the upper right direction (dotted line in Fig. 1). There are no clear dominance relationships among the brands, but there are some. We focus our attention to the four categories generated by combining the two characteristics; cola/lemon lime and diet/non-diet, and examine the dominance relationship in each category. Each category has two brands. It is possible to say which brand is dominant over the other brand. Along Dimension 1, Coke and Pepsi are almost on the 45° line. It seems that the two brands have almost the same dominance. 7-Up seems dominant over Sprite. Diet Pepsi is dominant over Tab, while the magnitude of the brand switching from Diet Pepsi to Tab as well as that from Tab to Diet Pepsi are small, because the outward and inward tendencies of the two brands are small. Like is dominant over Fresca. The magnitude of the brand switching from Like to Fresca and that from Fresca to Like are small by the same reason as in the case between Diet Pepsi and Tab.

Along Dimension 2, Pepsi and 7-Up in the second quadrant are dominant over the five brands in the first quadrant. Coke in the third quadrant is dominant over Pepsi and 7-Up. Like and Fresca (lemon-lime diet brands) are dominant over the other three brands in the first quadrant. While there is not any significant dominant relationships among brands along Dimension 3, Coke is dominant over the other three diet brands in the third quadrant. Like (diet lemon-lime brand) in the fourth quadrant is dominant over the four brands in the third quadrant. And three brands (Pepsi, 7-Up and Sprite) which are non-diet brands in the first quadrant are dominant over Like. Along Dimension 4, there are no significant dominance relationships among brands in the first and the third quadrants respectively. Like in the second quadrant is dominant over the two brands (non-diet lemon-lime brands) in the first quadrant, and is dominated by the four brands in the third quadrant (which include three cola brands). Tab (diet cola brand) in the fourth quadrant is dominant over the four brands (three are cola brands) in the third quadrant. 7-Up and Pepsi in the first quadrant are dominant over Tab.

It has been disclosed that the difference between diet and non-diet brands plays an important role in the brand switching among the eight soft drink brands (Bass et al. 1972; DeSarbo and Soete 1984). This was also confirmed in the present study. It also disclosed that (a) the difference between cola and lemon-lime brands also plays an important role in the brand switching as well, and that (b) Coke and Pepsi play an important role in the brand switching, respectively.

In the present asymmetric multidimensional scaling the diagonal element of the brand switching matrix was dealt with. The diagonal element represents the magnitude of staying or the loyalty to the corresponding brand. The reproduced diagonal element of brand j based on four dimensions is $d_1 x_{j1} y_{j1} + d_2 x_{j2} y_{j2} + d_3 x_{j3} y_{j3} + d_4 x_{j4} y_{j4}$. The reproduced diagonal element correlates with the diagonal element of the brand switching matrix positively ($r = 0.94$), with the market share (Bass et al. 1972) positively ($r = 0.97$), and with the density (DeSarbo and Soete 1984) which tells the dominance of the brand (the smaller density means the larger dominance) negatively ($r = -0.80$). These figures tell that the outward tendency and the inward tendency of a brand represent the magnitude of staying or the loyalty to the brand.

Acknowledgements The author would like to express his gratitude to two anonymous referees who gave the author valuable suggestions to the earlier version of the paper.

References

Bass FM, Pessemier EA, Lehmann DR (1972) An experimental study of relationships between attitudes, brand preference, and choice. Behav Sci 17(6):532–541

Borg I, Groenen PJF (2005) Modern multidimensional scaling: theory and applications, 2nd edn. Springer, New York

DeSarbo WS, Soete GD (1984) On the use of hierarchical clustering for the analysis of nonsymmetric proximities. J Consum Res 11(1):601–610

DeSarbo WS, Manrai AK, Burke RR (1990) A nonspatial methodology for the analysis of two-way proximity data incorporating the distance-density hypothesis. Psychometrika 55(2):229–253

Kleinberg JM (1999) Authoritative sources in a hyperlinked environment. J ACM 46(5):604–632

Okada A (2008) Two-dimensional centrality of a social network. In: Preisach C, Burkhardt H, Schmidt-Thieme L, Decker R (eds) Data analysis, machine learning and applications, studies in classification, data analysis, and knowledge organization. Springer, Heidelberg, pp 381–388

Okada A (2011) Centrality of asymmetric social network: singular value decomposition, conjoint measurement, and asymmetric multidimensional scaling. In: Ingrassia S, Rocci R, Vichi M (eds) New perspectives in statistical modeling and data analysis, studies in Classification, data analysis, and knowledge organization. Springer, Heidelberg, pp 219–227

Okada A, Tsurumi H (2012) Asymmetric multidimensional scaling of brand switching among margarine brands. Behaviormetrika 39:111–126

Zielman B, Heiser WJ (1991) Analysis of asymmetry by a slide vector. Technical Report RR91-05, Department of Data Theory, University of Leiden, Leiden

Automatic Regularization of Factorization Models

Steffen Rendle

Abstract Many recent machine learning approaches for prediction problems over categorical variables are based on factorization models, e.g. matrix or tensor factorization. Due to the large number of model parameters, factorization models are prone to overfitting and typically Gaussian priors are applied for regularization. Finding proper values for the regularization parameters is usually done with an expensive grid-search using holdout validation data. In this work, two approaches are presented where regularization values are found without increasing computational complexity. The first one is based on interweaving optimization of model parameters and regularization in stochastic gradient descent algorithms. Secondly, a two-level Bayesian model to integrate regularization values into inference is shortly discussed.

1 Introduction

Recently, factorization models such as matrix or tensor factorization models have attracted a lot of research due to their success in important applications, e.g. in recommender systems and the Netflix prize.[1] Many different factorization models have been proposed where a common characteristic is that all of them deal with a very large number of model parameters – especially if the rank of the factorization is chosen large. The large number of model parameters makes factorization approaches prone to overfitting. To achieve a high generalization quality,

---

[1] http://www.netflixprize.com/

S. Rendle (✉)
University of Konstanz, Konstanz, Germany
e-mail: steffen.rendle@uni-konstanz.de

W. Gaul et al. (eds.), *German-Japanese Interchange of Data Analysis Results*,
Studies in Classification, Data Analysis, and Knowledge Organization,
DOI 10.1007/978-3-319-01264-3_14,

regularization is applied. Such regularized factorization models have shown great prediction quality – provided that the regularization constants have been chosen correctly. However, the selection of regularization values is typically done with grid-search approaches which are very time-consuming and only applicable when the number of regularization parameters is low. In recent research (Rendle 2012), a simple extension for stochastic gradient descent (SGD) based learning algorithms has been proposed that finds regularization values while model parameters are learned. In this work, this approach is generalized for grouping of variables and extended to other regularization functions.

2 Regularized Factorization Models

In the following, the problem setting is defined in a generic way that subsumes many of the state-of-the-art factorization models and their regularization structures.

2.1 Factorization Models

The most well-studied factorization model is matrix factorization (MF) (e.g. Srebro et al. 2005) where the interaction between two categorical variables C_1 and C_2 is modeled with the dot product of two latent vectors:

$$\hat{y}(c_1, c_2) := \langle \mathbf{v}_{c_1}, \mathbf{v}_{c_2} \rangle = \sum_{f=1}^{k} v_{c_1,f}\, v_{c_2,f}, \quad V \in \mathbb{R}^{C_1 \cup C_2} \tag{1}$$

Here V are the model parameters that should be learned – i.e. each level c of the categorical variables is represented by a k dimensional vector $\mathbf{v}_c$. There are many extensions for more complex predictor variables, e.g. tensor factorization models (e.g. Tucker 1966) for more than two categorical variables or specialized models that take other types of variables into account, e.g. implicit models and neighborhood information (e.g. Koren 2010).

Factorization machines (FM) (Rendle 2010) are an attempt to generalize factorization models by using design matrices with p real-valued predictor variables, $\mathbf{x} \in \mathbb{R}^p$, as input. This makes FMs as flexible as standard machine learning models, e.g. linear/polynomial regression or support vector machines (SVMs). An FM models all nested interactions up to order d between the p input variables in $\mathbf{x}$ using factorized interaction parameters. The factorization machine (FM) model of order $d = 2$ is defined as

$$\hat{y}(\mathbf{x}) := w_0 + \sum_{j=1}^{p} w_j\, x_j + \sum_{j=1}^{p} \sum_{j'=j+1}^{p} x_j\, x_{j'} \sum_{f=1}^{k} v_{j,f}\, v_{j',f} \tag{2}$$

where k is the dimensionality of the factorization and the model parameters $\Theta = \{w_0, w_1, \ldots, w_p, v_{1,1}, \ldots v_{p,k}\}$ are

$$w_0 \in \mathbb{R}, \quad \mathbf{w} \in \mathbb{R}^p, \quad V \in \mathbb{R}^{p \times k}. \tag{3}$$

Like polynomial regression, an FM includes all interactions between variables up to order d. In Eq. (2), the first part corresponds to linear regression (unary interactions) and the second part contains all interactions between variable pairs – e.g. $x_j\, x_{j'}$ is the interaction between the j-th and j'-th variable. However, instead of using one independent model parameter per interaction (e.g. $w_{j,j'}$ for the interaction between x_j and $x_{j'}$), FMs use a low rank assumption (e.g. the pairwise effect between x_j and $x_{j'}$ is modeled by $\sum_{f=1}^{k} v_{j,f}\, v_{j',f}$). This correspond to the assumption that the matrix of all p^2 pairwise effects $W \approx V\, V^t$ has a low rank $k \ll p$. This reduces the number of model parameters for pairwise effects from p^2 to $p\,k$ and allows to estimate interactions even in problems where the number p of variables is very large.

2.2 Regularization

Factorization models are typically applied for prediction problems with a large number of predictor variables, i.e. p is large – often in the range of millions. This includes matrix factorization which is the most common factorization model where the predictor variables can be seen as the levels of two categorical variables of large domain. In contrast to standard linear regression models that have $\mathcal{O}(p)$ model parameters Θ, this number increases to $\mathcal{O}(k\,p)$ for factorization models, where a typical value for k is 100. For example, a simple matrix factorization model for the Netflix prize data could have $k \times p = 100 \times 500{,}000 = 50{,}000{,}000$ model parameters that are fitted to 100,000,000 observations. This high number of model parameters makes factorization models prone to overfitting.

In all state-of-the art factorization approaches, a regularization function R is applied to prevent overfitting. The most common approach of regularization is to favor small values for model parameters Θ because the lower the value the less the effect of the model parameter on the outcome and thus the lower the complexity of the solution. For example, in L2 regularization a quadratic penalty (i.e. $R(\theta) = \theta^2$) is used which corresponds from a probabilistic view to the assumption that a model parameter follows a Gaussian prior $\theta \sim \mathcal{N}(0, 1)$. The regularizers R themselves typically depend on Π many regularization parameters $\Lambda = \{\lambda_1, \ldots, \lambda_\Pi\}$ which can be used e.g. for specifying the strength of the regularization (e.g. $R(\theta) = \lambda \theta^2$ or analogously $\theta \sim \mathcal{N}(0, 1/\lambda)$).

2.2.1 Grouping of Model Parameters

In many factorization models, there is more than one regularization parameter (i.e. $\Pi > 1$) so that different parts of the model can be regularized with individual constants. E.g. model parameters for pairwise and unary effects (see Eq. (2)) might need different regularization constants. In this work, this general setting is studied by using a partition function $\pi : \Theta \rightarrow \{1, \ldots, \Pi\}$ that allows to place each model parameter in one of Π groups. The partition function π is the link between the Π regularization variables Λ and the model parameters Θ – e.g. the regularization parameter for model parameter $\theta \in \Theta$ is $\lambda_{\pi(\theta)}$. And the set of model parameters that share the l-th group (and thus also share $\lambda_l \in \Lambda$) can be denoted by $\pi^{-1}(l) \subseteq \Theta$.

Some choices for partition functions are:

1. No grouping:

$$\Pi = 1, \quad \pi^{-1}(1) = \Theta, \tag{4}$$

2. Individual regularization parameters per factor layer:

$$\Pi = 2 + k, \quad \pi^{-1}(1) = \{w_0\}, \tag{5}$$

$$\pi^{-1}(2) = \{w_i : i \in \{1, \ldots, p\}\}, \tag{6}$$

$$\pi^{-1}(2 + f) = \{v_{i,f} : i \in \{1, \ldots, p\}\}, \ \forall f \in \{1, \ldots, k\}. \tag{7}$$

3. Depending on the application, it also makes sense to divide the predictor variables into groups, e.g. the first variables within $\mathbf{x}$ might describe the levels of one categorical variable and thus can be modeled with an individual regularization parameter. E.g. if $\mathbf{x} \in \mathbb{R}^{2,000}$ describes two categorical variables with each 1,000 levels, one reasonable grouping would be:

$$\Pi = 1 + 2, \quad \pi^{-1}(1) = \{w_0\}, \tag{8}$$

$$\pi^{-1}(2) = \{w_i, v_{i,f} : i \in \{1, \ldots, 1,000\}, f \in \{1, \ldots, k\}\}, \tag{9}$$

$$\pi^{-1}(3) = \{w_i, v_{i,f} : i \in \{1,001, \ldots, 2,000\}, f \in \{1, \ldots, k\}\}. \tag{10}$$

This grouping can also be combined with the regularization per factor layer.

2.2.2 Regularization Functions

In this work, regularizers of the following form are studied:

$$R(\Theta, \Lambda) = \sum_{g \in \{1, \ldots, \Pi\}} \lambda_g \, r(\pi^{-1}(g)) \tag{11}$$

where r is a real-valued function over a subset of model parameters (selected by the grouping). Examples for r are the L2-norm which corresponds to Gaussian priors on the model parameters[2] or the L1-norm corresponding to Laplace priors

$$r^{L2}(\Theta') = \sum_{\theta \in \Theta'} \theta^2, \quad r^{L1}(\Theta') = \sum_{\theta \in \Theta'} |\theta_i|. \tag{12}$$

Here Θ' is a subset of model parameters, $\Theta' \subseteq \Theta$.

From an optimization perspective, L2 punishes large values of model parameters with a quadratic penalty. With L1 regularization, sparse solutions can be achieved which can be seen as a kind of feature selection (e.g. Zou and Hastie 2005).

2.3 Learning Regularized Factorization Models

Similar to other statistical models, the parameters of factorization models are learned using observed data. Let $S = \{(\mathbf{x}_1, y_1), (\mathbf{x}_2, y_2), \ldots\}$ be the set of observed training instances. The loss L measures how well the model parameters explain the training data

$$L(S, \Theta) = \sum_{(\mathbf{x}, y) \in S} l(\hat{y}(\mathbf{x}), y) \tag{13}$$

where l is a loss function, e.g. squared loss $l^{LS}(y_1, y_2) = (y_1 - y_2)^2$ for regression or logit loss $l^C(y_1, y_2) = \ln(1 + \exp(-y_1 y_2))$ for classification.

The overall optimization criterion $OptReg$ for model parameters Θ given regularization constants Λ is a combination of loss L and regularization R

$$\Theta^* | \Lambda := \text{OptReg}(S, \Lambda) := \underset{\Theta}{\text{argmin}} \left(L(S, \Theta) + R(\Theta, \Lambda) \right) \tag{14}$$

Here $\Theta^* | \Lambda$ is used to emphasize that Θ^* is optimal only w.r.t. given regularization constants Λ. To simplify notation, only one regularization function R is used here, but this could also be extended for a linear combination of several functions, e.g. combining L1 and L2 regularization would correspond to an elastic net (Zou and Hastie 2005). The adaptive SGD algorithm described in Sect. 3.1 can handle several functions R.

---

[2]The maximum a posteriori estimator of a likelihood with Gaussian priors is equivalent to a loss with L2 regularization where the regularization constant λ corresponds to the precision (or inverse variance) of the Gaussian prior.

2.3.1 Learning Algorithm

Stochastic gradient descent (SGD) is the most common technique for learning factorization models. In contrast to full/batch gradient descent (GD) which computes the complete gradient of the optimization criterion for each step, SGD performs a small gradient step for each single training case. Compared to other optimization techniques such as alternating least-squares or coordinate descent, SGD is (i) very easy to implement and (ii) easy to adapt to a wide variety of loss functions.

SGD algorithms iterate in a random order over all training examples and for each single training example $(\mathbf{x}, y) \in S$, a small update step (with step size η) is made in the opposite direction of the gradient of the objective (Eq. 14)

$$\theta^{t+1} = \theta^t - \eta \left(\frac{\partial}{\partial \theta} l(\hat{y}(\mathbf{x}), y) + \lambda_{\pi(\theta)} \frac{\partial}{\partial \theta} r(\pi^{-1}(\pi(\theta))) \right) \tag{15}$$

where θ^t is the model parameter θ after t gradient steps. The derivation of the losses is for square loss

$$\frac{\partial}{\partial \theta} l^{LS}(\hat{y}(\mathbf{x}|\Theta), y) = \frac{\partial}{\partial \theta} (\hat{y}(\mathbf{x}|\Theta) - y)^2 = 2 (\hat{y}(\mathbf{x}|\Theta) - y) \frac{\partial}{\partial \theta} \hat{y}(\mathbf{x}|\Theta) \tag{16}$$

and for logit loss (using the definition of $\sigma(z) = 1/(1 + \exp(-z))$)

$$\frac{\partial}{\partial \theta} l^{C}(\hat{y}(\mathbf{x}|\Theta), y) = \frac{\partial}{\partial \theta} \ln(1 + \exp(-\hat{y}(\mathbf{x}|\Theta) y)) = (\sigma (\hat{y}(\mathbf{x}|\Theta) y) - 1) y \frac{\partial}{\partial \theta} \hat{y}(\mathbf{x}|\Theta). \tag{17}$$

The derivation of the regularization function for L2 and L1 is

$$\frac{\partial}{\partial \theta} r^{L2}(\{\theta_1, \ldots, \theta_l\}) = 2 \theta, \quad \frac{\partial}{\partial \theta} r^{L1}(\{\theta_1, \ldots, \theta_l\}) = \begin{cases} 1, & \text{if } \theta > 0, \\ -1, & \text{if } \theta < 0, \\ 0, & \text{else} \end{cases} \tag{18}$$

The convergence of SGD depends on the choice of the step size η which has to be chosen 'small enough' for the specific data and regularization setting. The step size is typically determined by a hyperparameter search.

3 Automatic Regularization

The SGD algorithm described so far is the standard learning technique for most regularized factorization models. In this algorithm, the regularization values are constants and have to be found in a model selection phase, which is done typically with expensive grid-search. In the following, two approaches are described where the search for regularization values is integrated in the learning algorithm for model parameters.

3.1 Adaptive SGD (SGDA)

In (Rendle 2012) an extension for SGD algorithms is proposed where regularization values are optimized while model parameters are learned. This approach is discussed in the following and generalized to grouping and non-L2 regularization.

3.1.1 Optimization Task for Regularization Values

To measure the quality of regularization parameters Λ, typically a holdout set is used. Let $S_V \subset S$ be a validation set, then

$$\Lambda^* := \operatorname*{argmin}_{\Lambda} \left(L(S_V, \text{OptReg}(S \setminus S_V, \Lambda)) \right) \qquad (19)$$

Note that Λ is optimized on the validation set and Θ on the disjoint set of remaining cases $S \setminus S_V$. Mostly this objective is solved by using candidates for Λ (e.g. grid search) and solving OptReg for each candidate of Λ independently. Clearly, this is a very time consuming approach and mostly infeasible for a large number of regularization values.

3.1.2 Integrated SGD Algorithm

In (Rendle 2012) learning regularization values is integrated into the SGD learning algorithm for model parameters. The approach is based on the observation that in each SGD-step on model parameters (Eq. 15) the regularization value has an influence on the future model parameter. Now, the regularization values can be adapted so that the next update on model parameters leads to a minimal error on the validation set. To formalize and apply this idea, the model equation (Eq. 2) after the next SGD-step on model parameters is defined as

$$\hat{y}(\mathbf{x}|\Theta^{t+1}) := w_0^{t+1} + \sum_{j=1}^{p} w_j^{t+1} x_j + \sum_{j=1}^{p} \sum_{j'=j+1}^{p} x_j x_{j'} \sum_{f=1}^{k} v_{j,f}^{t+1} v_{j',f}^{t+1} \qquad (20)$$

where each θ^{t+1} is defined according to Eq. (15), i.e. it depends explicitly on Λ.

With this definition, the validation set error can be minimized on S_V with SGD as well:

$$\lambda_g^{t+1} = \lambda_g^t - \eta \left(\frac{\partial}{\partial \lambda_g} l(\hat{y}(\mathbf{x}|\Theta^{t+1}), y) \right) \tag{21}$$

with

$$\frac{\partial}{\partial \lambda_g} \hat{y}(\mathbf{x}|\Theta^{t+1}) = -\eta \left[\delta(\pi(w_0) = g) \frac{\partial}{\partial w_0} r(\pi^{-1}(g)) + \sum_{j : \pi(w_j) = g} \frac{\partial}{\partial w_j} r(\pi^{-1}(g)) \right.$$
$$\left. + \sum_{f=1}^{k} \sum_{j : \pi(v_{j,f}) = g} \sum_{j' \neq j}^{p} x_j x_{j'} v_{j',f}^{t+1} \frac{\partial}{\partial v_{j,f}} r(\pi^{-1}(g)) \right] \tag{22}$$

Now a simple integrated SGD learning algorithm that optimizes both model and regularization parameters jointly can be sketched. The SGDA algorithm alternates between updating the model parameters with Eq. (15) based on one case of $S \setminus S_V$ and updating regularization parameters with Eq. (21) based on one case of S_V. The overall computational complexity of this integrated algorithm is the same as learning only model parameters (Sect. 2.3.1).

3.2 Hierarchical Bayesian Model

A search for regularization values can also be avoided with hierarchical Bayesian models. From a Bayesian point of view, regularization can be seen as prior distributions over model parameters, e.g. L2 regularization with λ is the same as a 0-mean Gaussian prior with variance $1/\lambda$. By defining a hyperprior distribution over the regularization values, e.g. a Gamma distribution, inference about the regularization values is possible:

$$\lambda_g \sim \Gamma(\alpha, \beta), \quad \lambda_g|\Theta, \alpha, \beta \sim \Gamma \left(\frac{\alpha + |\pi^{-1}(g)| + 1}{2}, \frac{\beta + \sum_{\theta \in \pi^{-1}(g)} \theta^2}{2} \right) \tag{23}$$

For factorization models, such an approach has been proposed for MF by Salakhutdinov and Mnih (2008) and for generic FMs by Freudenthaler et al. (2011). Equation (23) extends the Bayesian FM approach of Freudenthaler et al. (2011) by groups.

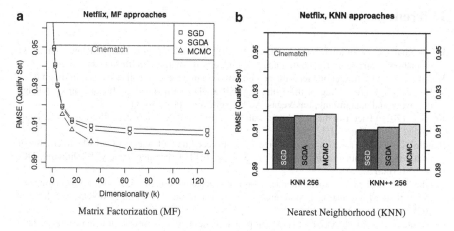

a Netflix, MF approaches   **b** Netflix, KNN approaches

Matrix Factorization (MF)   Nearest Neighborhood (KNN)

Fig. 1 Comparison of the automatic regularization approaches SGDA and MCMC to the expensive SGD approach. (**a**) Depicts the error of a matrix factorization model (with varying k) and (**b**) for the KNN approaches of Koren (2008). Results are from Rendle (2012) and Freudenthaler et al. (2011)

4 Evaluation

Figure 1 compares standard SGD learning to the automatic regularization approaches of adaptive SGD (*SGDA*, Sect. 3.1), and the Hierarchical Bayesian Model (*MCMC*, Sect. 3.2). Two recommender system models are shown, matrix factorization (left) and nearest neighbor (right) both regularized with L2. All experiments[3] are conducted on the Netflix prize dataset with about 100,000,000 training cases and 1,400,000 test cases.

For the MF models, MCMC has a lower error than SGD and SGDA. The reason is that SGD and SGDA are point estimators, whereas MCMC estimates the whole predictive distribution by sampling and thus can cover uncertainty.

5 Conclusion

Regularization is a an important aspect in learning factorization models because of their high number of model parameters. This work discusses two possibilities to find proper regularization values without increasing the computational complexity of the learning algorithms even when the number of regularization parameters is high.

---

[3] All models have been learned with LIBFM, http://www.libfm.org.

References

Freudenthaler C, Schmidt-Thieme L, Rendle S (2011) Bayesian factorization machines. In: NIPS workshop on sparse representation and low-rank approximation, Sierra Nevada

Koren Y (2008) Factorization meets the neighborhood: a multifaceted collaborative filtering model. In: KDD '08: proceeding of the 14th ACM SIGKDD international conference on knowledge discovery and data mining, Las Vegas. ACM, New York, pp 426–434

Koren Y (2010) Factor in the neighbors: scalable and accurate collaborative filtering. ACM Trans Knowl Discov Data 4:1:1–1:24

Rendle S (2010) Factorization machines. In: Proceedings of the 2010 IEEE international conference on data mining, ICDM '10, Sydney. IEEE Computer Society, Washington, DC, pp 995–1000

Rendle S (2012) Learning recommender systems with adaptive regularization. In: Proceedings of the fifth ACM international conference on web search and data mining, WSDM '12, Seattle. ACM, New York, pp 133–142

Salakhutdinov R, Mnih A (2008) Bayesian probabilistic matrix factorization using Markov chain Monte Carlo. In: Proceedings of the 25th international conference on machine learning, ICML '08, Helsinki. ACM, New York, pp 880–887

Srebro N, Rennie JDM, Jaakkola TS (2005) Maximum-margin matrix factorization. In: Saul LK, Weiss Y, Bottou L (eds) Advances in neural information processing systems 17, Vancouver. MIT, Cambridge, pp 1329–1336

Tucker LR (1966) Some mathematical notes on three-mode factor analysis. Psychometrika 31:279–311

Zou H, Hastie T (2005) Regularization and variable selection via the elastic net. J R Stat Soc B (Stat Methodol) 67(2):301–320

Three-Way Data Analysis for Multivariate Spatial Time Series

Mitsuhiro Tsuji, Hiroshi Kageyama, and Toshio Shimokawa

Abstract We discuss several methods to realize three-way (three mode) approaches to clustering using the INDCLUS model and multidimensional scaling using the INDSCAL model, which assumes that the objects are embedded in a discrete or continuous space common to all data, including individual differences obtained by weighting each dimension. We apply some effective dynamic graphical approaches using two methods to perform a time-space structural analysis for multivariate spatial time series. The clustering and scaling of multivariate spatial time series consider: (1) the spatial nature of the objects to be clustered geometrically (discrete); (2) the characteristics of the feature space with the time series (continuous); (3) the latent structure between space and time. The last aspect is addressed using dynamic graphics with a matrix-type presentation. We can simultaneously observe the spatial nature, move the feature space and can zoom in/out of the results using a suitable size. The proposed analysis can be applied to the classification and scaling of the prefectures of Japan on the basis of the observed dynamics of some safety indicators.

1 Spatial Time Series Data

The statistical analysis of spatial time series data, which include the place and time of collection, is very complex. Uncertainty and the role of statistics should be considered in detail (Cressie and Wikle 2011). Coppi, D'urso and Giordani

M. Tsuji (✉) · H. Kageyama
Kansai University, Takatsuki, Osaka, Japan
e-mail: tsuji@kansai-u.ac.jp

T. Shimokawa
University of Yamanashi, Kofu, Yamanashi, Japan
e-mail: shimokawa@yamanashi.ac.jp

W. Gaul et al. (eds.), *German-Japanese Interchange of Data Analysis Results*,
Studies in Classification, Data Analysis, and Knowledge Organization,
DOI 10.1007/978-3-319-01264-3_15,
© Springer International Publishing Switzerland 2014

considered the space of multivariate time trajectories and the spatial nature of clustering (Coppi et al. 2010).

In this report, we apply both the discrete INDCLUS and continuous INDSCAL approaches to some liquor sales data. We consider both the geographical space of liquor sales and the feature space of the time series at the same stage.

2 Three-Way Data Analysis

We propose and investigate three-way clustering and scaling approaches (Pruzansky 1975; Arabie et al. 1987; Tsuji et al. 2010) to spatial time series data. The data that we are investigating are three-way data, namely,

$$liquor \times time \times place$$

The three variables are characterized as follows:

- On the basis of five type of liquor (sake, white distilled liquor, beer, whisky, wine)
- On the basis of the year (spanning 1987–2007)
- On the basis of the location from among 46 prefectures (excluding Okinawa) in Japan,

in relation to safety problems of drunken driving. When driving in a bad condition that normal driving may be impaired by alcohol, the penalties to drivers have been strengthened. Data for liquor sales were converted into sales per 100 persons.

When we applied the INDCLUS and INDSCAL, one of the most difficult problems encountered was the selection of the initial configuration. Therefore, we focused on the continuous feature of time as follows:

- From 1987 to 2007, we calculated the mean of each liquor over 5 years (1987–1991, 1992–1996, 1997–2001, 2002–2006) to avoid the difficulty of choosing an initial configuration in the three-way data analysis INDCLUS and INDSCAL.

  Our research used the following procedure;

- First, we used the INDCLUS to explain the structure of places using Japanese maps.
- Next, we used the INDSCAL to explain the geometrical interpretation of places (prefectures) from the view point of individual weights of liquor and time.
- For the annual data from 1987 to 2007, we again used the INDSCAL with only one iteration to explain the geometrical interpretation of places (prefectures) from the view point of the individual weights of liquor and time.
- Finally, from the results of the individual weights, we were able to find the time trajectories.

Fig. 1 Three-cluster map using INDCLUS (VAF = 0.488)

2.1 Results by INDCLUS

The INDCLUS model is represented by

$$s_{ij,k} \cong \sum_{r=1}^{R} w_{kr} p_{ir} p_{jr} + c_k,$$

where $s_{ij,k}$ is the similarity between place i and j for individual k ($k = 1, \cdots, K$); R is the suitable number of clusters found from places; w_{kr} is the weight of individual liquor k for cluster r, p_{ir} represents the probability that place i belongs to cluster r ($p_{ir} = 1$) or not ($p_{ir} = 0$), and c_k is an additive constant of individual liquor k.

Characteristics of the INDCLUS model are overlapping clustering and non-hierarchical structure. Figure 1 shows the result map obtained using this model (VAF = 0.488) in the case of three clusters.

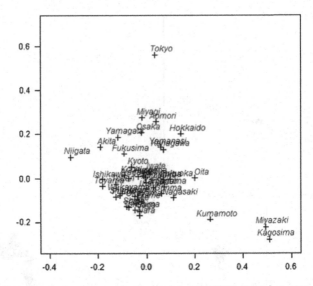

Fig. 2 Two-dimensional plot of places (prefectures) by INDSCAL (VAF = 0.464)

- The first cluster includes Yamanashi (wine) and Tokyo (beer).
- The second cluster includes Niigata (sake) and Southern Kyushu (white distilloed liquor).
- The third cluster includes many prefectures.

Thus, the spatial nature of liquor could be expressed geometrically and discretely.

2.2 Results by INDSCAL

In the INDSCAL model, the three-way matrix $D \equiv \{d_{ij,k}\}$ is estimated from an input data matrix $\Delta \equiv \{\delta_{ij,k}\}$, where $d_{ij,k}$ is the distance between place i and j for individual k of liquor and time, and $\delta_{ij,k}$ is the dissimilarity between place i and j for individual k of liquor and time. Distance $d_{ij,k}$ is represented by

$$d_{ij,k} = \sqrt{\sum_{r=1}^{R} w_{kr} \left(x_{ir} - x_{jr}\right)^2},$$

where w_{kr} is the weight of importance along dimension r $(r = 1, \cdots, R)$ for individual k $(k = 1, \cdots)$; R is the number of dimensions; x_{ir} and x_{jr} are coordinates of the place i and j, respectively, along dimension r of an R-dimension with individual weights of liquor and time space (Fig. 2)

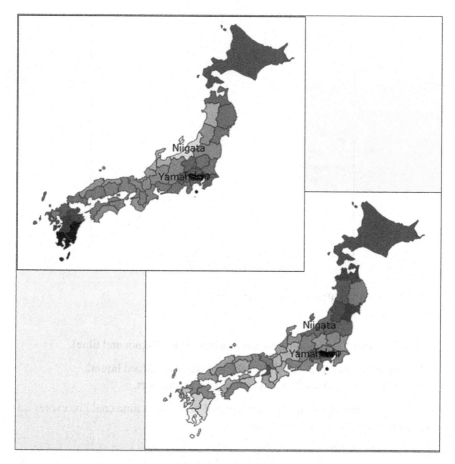

Fig. 3 Japan map of the results by INDSCAL with two-dimensions (VAF = 0.464)

- Along Axis-1, Niigata (sake) is located on the left-side and Kagoshima and Kumamoto (white distilled liquor) are located on the right-side.
- Along the Axis-2, Tokyo (beer) is located on the upper-side.

We can express the results for INDSCAL using a map of Japan (Fig. 3), where we can find the results for INDSCAL in a more geometrically continuous map.

- In the upper map, Niigata (sake) is represented by white, and Kagoshima and Kumamoto (white distilled liquor) are represented by deep black.
- In the lower map, Tokyo (beer) is represented by deep black.
- Several prefectures are identified by their shades.

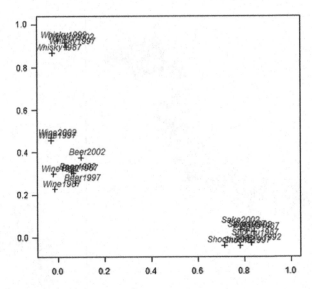

Fig. 4 Two-dimensional plot of weights (liquor and time)

Figure 4 shows the two-dimensional plot of weights (liquor and time).

- The weights of the Axis-1 include sake and white distilloed liquor.
- The weights of the Axis-2 include whisky, wine and beer.

Thus, the characteristics of the feature space of liquor with time could be expressed using a geometrically continuous map.

2.3   Individual Plot from Results of INDSCAL

In Sect. 2.2, we examined three time periods (1987–1991, 1992–1996, 1997–2001) to simplify the analysis of INDSCAL. To make the time series feature space, we create time series projects for the individual weight space.

Figure 5 shows a detailed two-dimensional plot of weights (liquor and time). In addition to that plot, we make the time trajectories for two weights of wine and sake.

We determined the continuity of time, and identified an interesting relationship between liquor and time. In 1993, the sale of wine changed significantly.

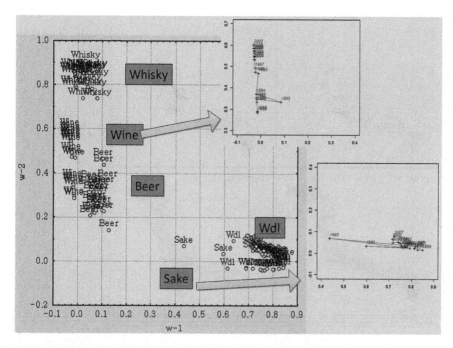

Fig. 5 Two-dimensional plot of weights (liquor and time) in detail

3 Dynamic Graphical Presentation with Matrix-Type

The validity of integration of INDCLUS (Cluster Analysis) and INDSCAL (MDS) was observed as follows;

- The result by INDCLUS was visualized geometrically.
- The result by INDSCAL was also visualized geometrically.

We developed the dynamic graphical presentation application with a matrix-type of some graphical files including geometrical results (Tsuji et al. 2009). Figure 6 shows the results of this application with a matrix-type of INDCLUS and INDSCAL. Then, we created various types of maps on the same plane, to show a more real configuration of the latent structure.

GIS (Geographical Information System) is a very interesting technology. Moreover, spatio-GIS includes some latent structure in the geographic space. The presentation of matrix-type dynamic graphics has some significant features, namely,

- We can obtain several results at the same time.
- We can simultaneously move several results into their appropriate places.
- We can zoom in/out of several results at the same time into their suitable size.

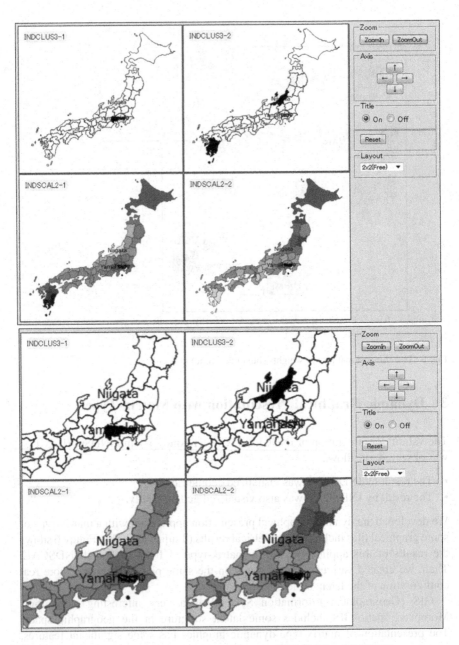

Fig. 6 Dynamic graphical presentation with matrix-type of INDCLUS and INDSCAL

Acknowledgements This work was supported by the MEXT-Supported Program for the Strategic
Research Foundation at Private Universities, 2008–2012.

References

Arabie P, Carroll JD, DeSarbo WS (1987) Three-way scaling and clustering. Sage, Newbury Park
Coppi R, D'Urso P, Giordani P (2010) A fuzzy clustering model for multivariate spatial time series.
  J Classif 27(1):54–88
Cressie N, Wikle CK (2011) Statistics for spatio-temporal data. Wiley, Hoboken
Pruzansky S (1975) How to use SINDSCAL: a computer program for individual differences in
  multidimensional scaling. Bell Telephone Laboratories, Murray Hill
Tsuji M, Konsha S, Shimokawa T (2009) A presentational approach to clustering results from
  regional structural analysis with time. In: CLADAG2009, Catania
Tsuji M, Shimokawa T, Okada A (2010) Three-way scaling and clustering approach to musical
  structural analysis. In: Locarek-Junge H, Weihs C (eds) Classification as a tool for research.
  Studies in classification, data analysis, and knowledge organization. Springer, Heidelberg,
  pp 767–774

Part III
Applications

Assessment of the Relationship Between Native Thoracic Aortic Curvature and Endoleak Formation After TEVAR Based on Linear Discriminant Analysis

Kuniyoshi Hayashi, Fumio Ishioka, Bhargav Raman, Daniel Y. Sze, Hiroshi Suito, Takuya Ueda, and Koji Kurihara

Abstract In the field of surgery treatment, thoracic endovascular aortic repair has recently gained popularity, but this treatment often causes an adverse clinical side effect called endoleak. The risk prediction of endoleak is essential for pre-operative planning (Nakatamari et al., J Vasc Interv Radiol 22(7):974–979, 2011). In this study, we focus on a quantitative curvature in the morphology of a patient's aorta, and predict the risk of endoleak formation through linear discriminant analysis. Here, we objectively evaluate the relationship between the side effect after stent-graft treatment for thoracic aneurysm and a patient's native thoracic aortic curvature. In addition, based on the sample influence function for the average of discriminant scores in linear discriminant analysis, we also perform statistical diagnostics on the result of the analysis. We detected the influential training

---

K. Hayashi (✉) · H. Suito · K. Kurihara
Graduate School of Environmental and Life Science, Okayama University, Okayama, Japan

CREST, Japan Science and Technology Agency, Tokyo, Japan
e-mail: k-hayashi@ems.okayama-u.ac.jp; suito@ems.okayama-u.ac.jp;
kurihara@ems.okayama-u.ac.jp

F. Ishioka
School of Law, Okayama University, Okayama, Japan

CREST, Japan Science and Technology Agency, Tokyo, Japan
e-mail: fishioka@law.okayama-u.ac.jp

B. Raman · D.Y. Sze
Department of Radiology, Stanford University School of Medicine, Stanford, USA
e-mail: ramanb@stanford.edu; dansze@stanford.edu

T. Ueda
Department of Radiology, St. Luke's International Hospital, Tokyo, Japan

CREST, Japan Science and Technology Agency, Tokyo, Japan
e-mail: takueda@luke.or.jp

W. Gaul et al. (eds.), *German-Japanese Interchange of Data Analysis Results*,
Studies in Classification, Data Analysis, and Knowledge Organization,
DOI 10.1007/978-3-319-01264-3_16,
© Springer International Publishing Switzerland 2014

samples to be deleted to realize improved prediction accuracy, and made subsets of all of their possible combinations. Furthermore, by considering the minimum misclassification rate based on leave-one-out cross-validation in Hastie et al. (The elements of statistical learning. Springer, New York, 2001, pp. 214–216) and the minimum number of training samples to be deleted, we deduced the subset to be excluded from training data when we develop the target classifier. From this study, we detected an important part of the native thoracic aorta in terms of risk prediction of endoleak occurrence, and identified influential patients for the result of the discrimination.

1 Introduction

Recently, many medical imaging tools have been developed, and they have provided quantitative and detailed data in the clinical field. In clinical treatment, doctors and researchers try to use the findings from these quantitative data for therapy planning. Therefore, statistical approaches and methods are useful for the development of high quality therapy planning.

Aneurysms are among diseases that are associated with aging and atherosclerotic change, and many people face the risk of developing an aneurysm. In the field of surgery treatment, thoracic endovascular aortic repair (TEVAR), which is a minimally-invasive therapy technique, has been widely accepted as a new treatment for thoracic aneurysms. Makaroun et al. (2008) compared the results of TEVAR and surgical repair in patients with descending thoracic aortic aneurysms, and they showed that TEVAR exhibited significantly lower incidences of aneurysm-related mortality. On the other hand, it has been shown that a clinical side effect called "endoleak" is often caused after TEVAR. Endoleak is one complication of TEVAR, and is a leakage of blood into the aneurysm sac after endovascular repair. Many investigators have recognized the contributions of aortic morphology and the achievement of adequate device fixation and seals for endoleak (Nakatamari et al. 2011). However, it seems that a quantitative analysis for the occurrence of endoleak has not been adequately performed. Then, Nakatamari et al. (2011) and Ishioka et al. (2011) applied linear discriminant analysis to the dataset that has quantitative variables computed on the basis of CT angiography, and they performed risk prediction. However, the normalization of the length of the thoracic aorta for each patient was not sufficiently processed. Moreover, statistical diagnostics of the analysis result were not performed in their studies.

In this paper, we properly normalized the length of the thoracic aorta of a patient, and we generated the target dataset for analysis. Then, we performed the discrimination of the no endoleak and endoleak groups in order to detect an important part of the thoracic aorta in the risk prediction of endoleak formation. In addition, we assessed the result of the discriminant analysis with the sample influence function (SIF), and detected notable patients in terms of the risk prediction of endoleak formation.

2   Data Set and Data Normalization

In one of the six FDA-sponsored clinical trials examining the Thoracic Excluder or TAG stent-graft devices (W.L. Gore & Associates., Flagstaff, Arizona), 121 patients were prospectively enrolled between April 2001 and September 2008 (Nakatamari et al. 2011). We preliminarily excluded the patients for whom pre-procedure CT angiography data were not available. In this study, we included 45 patients (no endoleak: 23 patients, endoleak: 22 patients). There are mainly four types of endoleak, but in this study, we performed a two-class discrimination between no endoleak and endoleak.

2.1   Calculation of Curvature Index

With reference to Rubin et al. (1998), we defined a curvature index $\kappa = 1/D$ cm^{-1} that quantifies aortic morphology, where D is the mean diameter. D can be calculated at 1-mm increments along the median luminal centerline computed based on CT angiography (see Fig. 1). In this study, we calculated the curvature index (κ) as the inverse of the radius of curvature at 10-mm discrete intervals along the aortic centerline. We regarded the part from the position of 30 points (30 mm) before the right brachio-cephalic artery to the celiac artery as the target part of the thoracic aorta being analyzed (see Fig. 2).

2.2   Normalization of the Length of the Target Part of the Thoracic Aorta

The target part of the thoracic aorta in each patient is different. Therefore, we need to normalize the length of this part. In this subsection, we describe how to normalize the target part in this study.

We first detected a patient (referred to as A38) who had the shortest length of the target part of all the patients. The length for patient A38 was 263 points. Next, we calculated the normalized points for all patients based on the length of the target part of A38. In addition, we estimated the curve line of the curvature data for each patient using B-splines of order three based on de Boor (1978) and Konishi and Kitagawa (2004, pp. 98–104). With the normalized position of the thoracic aorta and the estimated curve line function, we finally calculated the normalized curvature for each patient. Then, their lengths were normalized into 263 points. In patient A38, the curvature data along the part from point 221 to point 263 corresponded to missing values. Therefore, we adopted the maximum values for each of the 11 points from points 1–220, as the variables in discriminant analysis. The first variable indicates the maximum curvature from points 1–11, and the 20th variable corresponds to

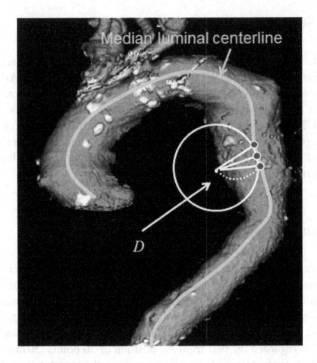

Fig. 1 An atherosclerotic aneurysm of a patient on a three-dimensional CT angiographic image

the maximum curvature from points 210–220. Moreover, we added the variables in terms of the initial and final positions for the part of stent-graft insertion on the normalized position. We performed the discrimination of no endoleak and endoleak with a dataset that had 45 training samples and 22 variables. The 21st and 22nd variables indicate the initial and final positions for the part of stent-graft insertion on the normalized position, respectively.

3 Linear Discriminant Analysis and Its Diagnostics

In this section, we give a simple explanation of linear discriminant analysis for two groups with reference to Hastie et al. (2001, pp. 84–88), and we also show statistical diagnostics based on the sample influence function with reference to Hampel et al. (1986, pp. 92–95), Fung (1992), and Tanaka (1994).

3.1 Overview of Linear Discriminant Analysis

Let X and G denote a p-dimensional real-valued random input vector and a categorical variable, respectively. In this paper, we code the classes of no endoleak

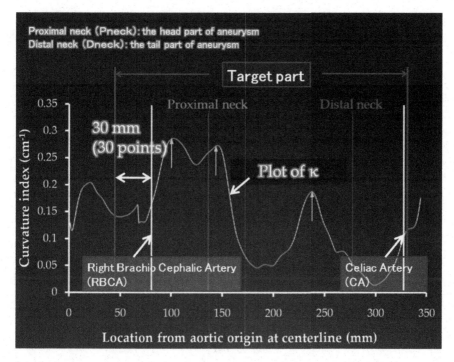

Fig. 2 The curvature of a patient as a function of location along the median centerline of the flow lumen

and endoleak as 1 and 2, respectively. Here, we assume the density functions of X in class 1 and class 2 to be $f_1(\mathbf{x})$ and $f_2(\mathbf{x})$, respectively. $\mathbf{x}$ is an input vector, and we assume the prior probability of a patient to be in class 1 and class 2 as π_1 and π_2, respectively, where $\pi_1 + \pi_2 = 1$.

By using $f_k(\mathbf{x})$ and $\pi_k (k = 1, 2)$,

$$P(G = k \mid X = \mathbf{x}) = \frac{P(X = \mathbf{x} \mid G = k)P(G = k)}{P(X = \mathbf{x})} = \frac{f_k(\mathbf{x})\pi_k}{f_1(\mathbf{x})\pi_1 + f_2(\mathbf{x})\pi_2} \quad (k = 1, 2). \tag{1}$$

In linear discriminant analysis, we assume the density function of class k to be $\frac{1}{(2\pi)^{p/2}|\Sigma|^{1/2}} \exp(-\frac{1}{2}(\mathbf{x} - \mu_k)^T \Sigma^{-1} (\mathbf{x} - \mu_k))$, where μ_k is the mean vector of the population and Σ is the common covariance matrix of the population. Based on (1) and $\frac{1}{(2\pi)^{p/2}|\Sigma|^{1/2}} \exp(-\frac{1}{2}(\mathbf{x} - \mu_k)^T \Sigma^{-1} (\mathbf{x} - \mu_k))$, $\log \frac{P(G=1|X=\mathbf{x})}{P(G=2|X=\mathbf{x})}$ can be written as

$$\mathbf{x}^T \Sigma^{-1}(\mu_1 - \mu_2) - \frac{1}{2}(\mu_1 + \mu_2)^T \Sigma^{-1}(\mu_1 - \mu_2) + \log \frac{\pi_1}{\pi_2}. \tag{2}$$

When formula (2) is greater than 0, we classify patient $\mathbf{x}$ into class 1. Otherwise, we classify $\mathbf{x}$ into class 2 based on Hastie et al. (2001, p. 88). In practical analysis, we use the estimated π_k, μ_k, and Σ that correspond to $\hat{\pi}_k = n_k/(n_1 + n_2)$, $\hat{\mu}_k = (1/n_k) \sum_{i=1}^{n_k} \mathbf{x}_i^k$, and $\hat{\Sigma} = (1/(n_1+n_2-2)) \sum_{k=1}^{2} \sum_{i=1}^{n_k} (\mathbf{x}_i^k - \hat{\mu}_k)(\mathbf{x}_i^k - \hat{\mu}_k)^{\mathrm{T}}$, respectively. Here, n_k is the number of training samples in the k-th class. $\mathbf{x}_i^k$ refers to the i-th training sample in the k-th class. With these estimated parameters, we calculate $\hat{f}_k(\mathbf{x}_i)$ and perform the discrimination in linear discriminant analysis.

3.2 Diagnostics Based on the Sample Influence Function

In this study, we define the average of the discriminant scores in each class as a target statistics because its statistical mean is the magnitude of the separation between classes. The discriminant score for $\mathbf{x}_i^1$ in class 1 is calculated as

$$\hat{z}_1(\mathbf{x}_i^1) = \log(\hat{f}_1(\mathbf{x}_i^1)\hat{\pi}_1) - \log(\hat{f}_2(\mathbf{x}_i^1)\hat{\pi}_2). \tag{3}$$

From (3), we can calculate the average of the discriminant scores in class 1 as follows.

$$\hat{Z}_1 = \frac{1}{n_1} \sum_{i=1}^{n_1} \hat{z}_1(\mathbf{x}_i^1). \tag{4}$$

The discriminant score for $\mathbf{x}_i^2$ in class 2 is

$$\hat{z}_2(\mathbf{x}_i^2) = \log(\hat{f}_2(\mathbf{x}_i^2)\hat{\pi}_2) - \log(\hat{f}_1(\mathbf{x}_i^2)\hat{\pi}_1). \tag{5}$$

From (5), we can also calculate the average of the discriminant scores in class 2 as

$$\hat{Z}_2 = \frac{1}{n_2} \sum_{i=1}^{n_2} \hat{z}_2(\mathbf{x}_i^2). \tag{6}$$

To assess the influence of the i-th training sample in the g-th class for $\hat{Z}_k$, we use the following measure as sample influence function (SIF):

$$\mathrm{SIF}(\mathbf{x}_i^g; \hat{Z}_k) = -(n_g - 1)(\hat{\tilde{Z}}_k - \hat{Z}_k), \tag{7}$$

where $\hat{\tilde{Z}}_k$ is the average of the discriminant scores in the k-th class when we develop the discriminant classifier or the discriminant rule by deleting $\mathbf{x}_i^g (i = 1,\ldots,n_g; g = 1, 2)$. We plot $\mathrm{SIF}(\mathbf{x}_i^g; \hat{Z}_k)$ along the i-th training sample of the g-th class to evaluate the influence of $\mathbf{x}_i^g$ for the k-th class. A positive value in the sign of $\mathrm{SIF}(\mathbf{x}_i^g; \hat{Z}_k)(k = 1, 2)$ indicates that the k-th class is not separated well from other classes. On the other hand, a negative value indicates that the k-th

Step 1	For the g-thc lass, detect the training samples that give at least one negative value for the $\mathrm{SIF}(\mathbf{x}_i^g; \hat{Z}_k)(k = 1, 2)$ forall $\hat{Z}_k(k = 1, 2)$.
Step 2	Calculate A_gs, which are all possible combinations of the training samples that give negative values in step 1 for the g-th class.
Step 3	For each A_g, develop the classifier by deleting A_g from the g-th training data, and calculate the misclassification rate of a leave-one-out cross-validation and a K-fold cross-validation.
Step 4	In each case, search the subsets that give the minimum misclassification rate, and find the subset that has the minimum number of training samples to be deleted among the subsets.
Step 5	Compare the number of training samples in the subset chosen based on a leave-one-out cross-validation and that based on a K-fold cross-validation.
Step 6	Choose the subset that has the minimum number of training samples of all the subsets in step 5.

Fig. 3 Diagnostic process based on the sample influence function

class is separated. We can, therefore, improve the predicted discriminant result by focusing on the negative sign. To assess the influence of multiple training samples on the analysis result, based on the training samples in the g-th class that give at least one negative value in $\mathrm{SIF}(\mathbf{x}_i^g; \hat{Z}_k)(k = 1, 2)$, we first calculate A_gs, which are all possible combinations of them in the g-th class. In each A_g, we then develop the classifier by deleting the subset A_g from the training data. We research the subsets that give the minimum misclassification rate in a leave-one-out cross-validation. Among them, we finally detect the subset that has the minimum number of training samples in the subsets. When we delete the influential training samples, we have to consider statistical efficiency. In general, we need to delete only the essential training samples from the perspective of prediction accuracy. Therefore, we should research the subsets that give the minimum misclassification rate based on the leave-one-out cross-validation and the K-fold cross-validation. Among them, we should detect the subset that has the minimum number of training samples. Then, by giving importance to statistical efficiency, we propose general diagnostics based in the sample influence function as shown in Fig. 3. In this study, we focused more on obtaining information about the relationship between the thoracic aortic morphology of a patient and the endoleak formation after TEVAR as opposed to the improvement of the prediction accuracy in discriminant analysis. Therefore, to perform an accurate diagnostics, in this paper, we performed only diagnostics that were based on the misclassification rate based on a leave-one-out cross-validation.

Table 1 Results

Predicted\observed	Endoleak (−)	Endoleak (+)	Total
Endoleak (−)	21	1	22
Endoleak (+)	2	21	23
Total	23	22	45

Table 2 Results of cross validation

Predicted\observed	Endoleak (−)	Endoleak (+)	Total
Endoleak (−)	13	8	21
Endoleak (+)	10	14	24
Total	23	22	45

Table 3 Results after variable selection (adopted variables: 1, 4, 6, 7, 10, 11, 13, 16, 18, 19, 21)

Predicted\observed	Endoleak (−)	Endoleak (+)	Total
Endoleak (−)	22	2	24
Endoleak (+)	1	20	21
Total	23	22	45

Table 4 Results of cross validation after variable selection (adopted variables: 1, 4, 6, 7, 10, 11, 13, 16, 18, 19, 21)

Predicted\observed	Endoleak (−)	Endoleak (+)	Total
Endoleak (−)	19	3	22
Endoleak (+)	4	19	23
Total	23	22	45

4 Risk Prediction of Endoleak Formation with Quantitative Data

We applied the discriminant rule in (2) to the target dataset described in Sect. 2.2. The results are shown in Tables 1–4.

4.1 General Results

Table 1 shows the results of a general discriminant analysis. The true discriminant rates in no endoleak and endoleak were 91.304 and 95.455 %, respectively. We identified a relationship between native curvature and endoleak formation, but these results based on a leave-one-out cross-validation were 56.522 and 63.636 %, respectively.

From these results in Table 2, in the case of 22 variables, we did not observe any significant relationship between native curvature and endoleak formation in terms of prediction.

Next, we performed variable selection using stepwise selection. In this variable selection process, we adopted the misclassification rate of a leave-one-out cross-validation. The general result and the result based on a leave-one-out cross-validation after variable selection are shown in Tables 3 and 4, respectively. The numbers on "adopted variables" in Tables 3 and 4 represent the chosen variables in the variable selection. In the former case, the true discriminant rates of no endoleak and endoleak were 95.652 and 90.909 %, respectively. In the latter case, they were 82.609 and 86.364 %, respectively. We determined that there were important parts among the positions on the native curvature that predicted the endoleak formation. Since the 21st variable chosen after variable selection is the variable that shows the initial position of stent-graft insertion, then, we confirmed the fact that the initial position is important for the prediction accuracy of the discrimination.

4.2   Statistical Diagnostics

The plots of $\mathrm{SIF}(\mathbf{x}_i^1; \hat{Z}_k)(i = 1, \ldots, 23; \ k = 1, 2)$ in Fig. 4 show the effects of $\mathbf{x}_i^1$ for $\hat{Z}_k$ when we develop the classifier by deleting $\mathbf{x}_i^1$ from class 1. In $\mathrm{SIF}(\mathbf{x}_i^1; \hat{Z}_1)$, there were six negative values to be deleted from the perspective of prediction. In $\mathrm{SIF}(\mathbf{x}_i^1; \hat{Z}_2)$, there were 14 negative values. The plots of $\mathrm{SIF}(\mathbf{x}_i^2; \hat{Z}_k)$ $(i = 1, \ldots, 22; \ k = 1, 2)$ in Fig. 5 show the effects of $\mathbf{x}_i^2$ for $\hat{Z}_k$ when we deleted the training samples in class 2. In $\mathrm{SIF}(\mathbf{x}_i^2; \hat{Z}_1)$, there were 16 negative values to be deleted. On the other hand, there were five negative values in $\mathrm{SIF}(\mathbf{x}_i^2; \hat{Z}_2)$. Therefore, in no endoleak and endoleak, there were the 17 training samples that give at least one negative value in the sample influence functions for the average discriminant scores in no endoleak and endoleak, respectively. Based on the misclassification rate in a leave-one-out cross-validation, we explored a minimum number of the subset that most improved the total accuracy of no endoleak and endoleak in the estimated discriminant model after variable selection. By the omission of the subset comprising the 1st, 2nd, 7th, 16th, 17th, 19th, 20th, 21st, and 23rd training samples in no endoleak, the total accuracy of no endoleak and endoleak based on a leave-one-out cross-validation was changed into 100 %. In the medical field, the cost of discriminant errors due to no endoleaks is different from that due to endoleaks. For example, based on Gore (2011, p. 10), it is mentioned that the percentage of the patients that have experienced an endoleak is 28.6 %. Then, we performed an additional analysis for multiple-case diagnostics. First, we explored the subsets to give the maximum total accuracy of endoleaks based on the misclassification rate in a leave-one-out cross-validation. Among these subsets, by considering statistical efficiency, we explored the best subset having the minimum number of deleted training samples, with the condition that the prediction accuracy of the no endoleak was equal or improved.

In the case of the perturbation at the training samples of no endoleak, there were two optimal subsets. One was the subset comprising the 1st, 2nd, 7th, 19th, 20th, 21st, and 23rd training samples in no endoleak. The other was the subset comprising the 1st, 2nd, 16th, 19th, 20th, 21st, and 23rd training samples in no endoleak. When we developed the classifier by the omission of the 1st, 2nd, 7th, 19th, 20th, 21st, and 23rd training samples in no endoleak, the accuracies in the general results for no endoleak and endoleak were 100 and 100 %, respectively. In addition, the

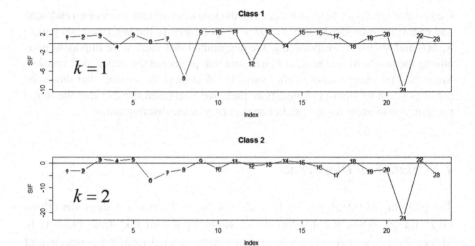

Fig. 4 The influence patterns obtained for the average of the discriminant scores in each class by deleting the training samples in no endoleak group

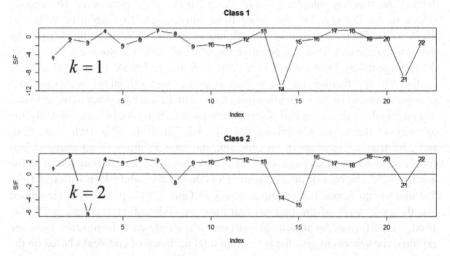

Fig. 5 The influence patterns obtained for the average of the discriminant scores in each class by deleting the training samples in endoleak group

accuracies in the results based on a leave-one-out cross-validation were 87.500 and 100 %, respectively. When we developed the classifier by the deletion of the 1st, 2nd, 16th, 19th, 20th, 21st, and 23rd training samples in no endoleak, the accuracies in the general results for no endoleak and endoleak were 93.75 and 100 %, respectively. Moreover, the accuracies in the results based on a leave-one-out cross-validation were 87.500 and 100 %, respectively. In the case of the perturbation at the training samples of endoleak, the subset comprising the 14th, 19th, and 21st training samples in endoleak was the best optimal subset. When we developed the classifier by excluding the 14th, 19th, and 21st training samples in endoleak, the true discriminant rates in the general results for no endoleak and endoleak were 95.652 and 100 %, respectively. In addition, the true discriminant rates based on a leave-one-out cross-validation were 86.957 and 100 %, respectively.

Therefore, the subset comprising the 1st, 2nd, 7th, 19th, 20th, 21st, and 23rd training samples or the 1st, 2nd, 16th, 19th, 20th, 21st, and 23rd training samples in no endoleak group was the optimal subset to delete from a point of view of prediction accuracy.

5   Discussion and Conclusion

Tse et al. (2004), Parmer et al. (2006), and Piffaretti et al. (2009) have investigated the association between the diameter and location of landing zones on a stent-graft, and the risk of endoleak formation after TEVAR, and they reported that the diameter and location are the most significant factors. Nakatamari et al. (2011) and Ishioka et al. (2011) have applied linear discriminant analysis to the dataset based on the native thoracic aortic curvature of a patient to study its effect for the risk of endoleak occurrence. However, an appropriate normalization for different lengths of thoracic aorta was not performed. In addition, statistical diagnostics for the analysis results were also not performed. In this study, we performed the discrimination for no endoleak and endoleak after appropriate processing. From the results based on a cross validation after variable selection, we detected the important parts (variables) of the thoracic aorta that indicate a risk of endoleak formation. In terms of risk prediction, in the statistical diagnostic analysis based on the sample influence function, we detected six notable patients in the no endoleak group. The 1st, 2nd, 19th, 20th, 21st, and 23rd training samples in no endoleak were chosen in the cases of the diagnostics based on the maximization of the total prediction accuracy and that of the prediction accuracy of endoleak group with the condition that the prediction accuracy of no endoleak group is equal or improved. In general, when we search the subsets to delete from the training data, if we focus only on improving the prediction accuracy of the classifier, we should search the best subsets for deleting not only an individual class, but all classes. However, in this study, we assigned higher priority to gaining an understanding of the characteristics of the patients in each group, and we searched each class.

In future, we will investigate their characteristics in detail. In this study, we also performed a statistical test for the difference between the two averages on the each variable chosen after variable selection, assuming that their population variances are different. From the results of these statistical tests shown in Tables 5–7, we determined that the curvature of patients in the no endoleak was larger than that in the endoleak in the 7th variable. On the other hand, we also observed the tendency whereby the curvature of patients in the endoleak was larger than that in

Table 5 Welch two sample t-test in each position after variable selection (by R: A language and environment for statistical computing, R Foundation for Statistical Computing, http://www.R-project.org/)

Position	No endoleak (average)	Endoleak (average)	p-value
Variable 1	21.171	21.430	0.852
Variable 4	19.438	18.649	0.679
Variable 6	23.577	22.034	0.490
Variable 7	27.900	22.654	0.022*
Variable 10	23.516	24.551	0.659
Variable 11	21.359	24.000	0.264
Variable 13	16.360	21.422	0.041*
Variable 16	12.161	16.529	0.037*
Variable 18	13.709	15.185	0.446
Variable 19	13.379	15.295	0.338
Variable 21	11.217	8.591	0.007**

Signif. code: 0.1;   0.05*;   0.01**;   0.001***

Table 6 Welch two sample t-test in each position in the case of deleting the 1st, 2nd, 7th, 16th, 17th, 19th, 20th, 21st, and 23rd training samples in no endoleak group after variable selection (by R: A language and environment for statistical computing)

Position	No endoleak (average)	Endoleak (average)	p-value
Variable 1	20.982	21.430	0.778
Variable 4	20.903	18.649	0.263
Variable 6	25.731	22.034	0.180
Variable 7	29.804	22.654	0.025*
Variable 10	21.041	24.551	0.154
Variable 11	19.470	24.000	0.063
Variable 13	15.771	21.422	0.047*
Variable 16	11.575	16.529	0.045*
Variable 18	12.462	15.185	0.213
Variable 19	13.240	15.295	0.323
Variable 21	12.429	8.591	0.004**

Signif. code: 0.1;   0.05*;   0.01**;   0.001***

the no endoleak in the 13th and 16th variables. In addition, in the 21st variable corresponding to the initial position of stent-graft insertion, we confirmed that the position in the no endoleak was farther from the origin of thoracic aorta than that in the endoleak. The rough averages of the initial positions of stent-graft insertion in the no endoleak and endoleak corresponded to farther positions than the 7th variable. Therefore, from the results in Tables 5–7, we could also confirm the tendency that the curvature value on the part of stent-graft insertion in the endoleak is larger than that in the no endoleak.

Table 7 Welch two sample t-test in each position in the case of deleting the 1st, 2nd, 19th, 20th, 21st, and 23rd training samples in no endoleak group after variable selection (by R: A language and environment for statistical computing)

Position	No endoleak (average)	Endoleak (average)	p-value
Variable 1	20.620	21.430	0.587
Variable 4	20.597	18.649	0.311
Variable 6	25.925	22.034	0.119
Variable 7	30.096	22.654	0.008**
Variable 10	21.706	24.551	0.227
Variable 11	19.378	24.000	0.046*
Variable 13	15.508	21.422	0.025*
Variable 16	11.048	16.529	0.015*
Variable 18	13.523	15.185	0.457
Variable 19	14.103	15.295	0.593
Variable 21	11.941	8.591	0.005**

Signif. code: 0.1; 0.05*; 0.01**; 0.001***

The findings in this study are useful for predicting the risk of endoleaks, and will support the therapy planning for TEVAR.

Acknowledgements This work was partly supported by the Core Research of Evolutional Science and Technology (CREST) in Japan Science and Technology Agency (Project: Alliance between Mathematics and Radiology).

References

de Boor C (1978) A practical guide to splines. Springer, New York

Fung WK (1992) Some diagnostic measures in discriminant analysis. Stat Probab Lett 13(4):279–285

Gore (2011) Gore tag thoracic endoprosthesis annual clinical update. August, p 10

Hampel FR, Ronchetti EM, Rousseeuw PJ, Stahel WA (1986) Robust statistics: the approach based on influence functions. Wiley, New York

Hastie T, Tibshirani R, Friedman J (2001) The elements of statistical learning. Springer, New York

Ishioka F, Nakatamari H, Suito H, Ueda T, Kurihara K (2011) Prediction of the future risk of endoleak complications based on statistical method. In: Proceedings 58th word statistics congress, International Statistical Institute, Dublin

Konishi S, Kitagawa G (2004) Information criterion. Asakurashoten, Tokyo, pp 98–104. (In Japanese)

Makaroun MS, Dillavou ED, Wheatley GH, Cambria RP (2008) Five-year results of endovascular treatment with the Gore TAG device compared with open repair of thoracic aortic aneurysms. J Vasc Surg 47(5):912–918

Nakatamari H, Ueda T, Ishioka F, Raman B, Kurihara K, Rubin GD, Ito H, Sze DY (2011) Discriminant analysis of native thoracic aortic curvature: risk prediction for endoleak formation after thoracic endovascular aortic repair. J Vasc Interv Radiol 22(7):974–979

Parmer SS, Carpenter JP, Stavropoulos SW, Fairman RM, Pochettino A, Woo EY, Moser GW, Bavaria JE (2006) Endoleaks after endovascular repair of thoracic aortic aneurysms. J Vasc Surg 44(3):447–452

Piffaretti G, Mariscalco G, Lomazzi C, Rivolta N, Riva F, Tozzi M, Carrafiello G, Bacuzzi A, Mangini M, Banach M, Castelli P (2009) Predictive factors for endoleaks after thoracic aortic aneurysm endograft repair. J Thorac Cardiovasc Surg 138(4):880–885

Rubin GD, Paik DS, Johnston PC, Napel S (1998) Measurement of the aorta and its branches with helical CT. Radiology 206(3):823–829

Tanaka Y (1994) Recent advance in sensitivity analysis in multivariate statistical methods. J Jpn Soc Comput Stat 7(1):1–25

Tse LW, MacKenzie KS, Montreuil B, Obrand DI, Steinmetz OK (2004) The proximal landing zone in endovascular repair of the thoracic aorta. Ann Vasc Surg 18(2):178–185

Fold Change Classifiers for the Analysis of Gene Expression Profiles

Ludwig Lausser and Hans A. Kestler

Abstract The classification of gene expression data is often based on profiles containing thousands of features. These features represent the abundance of RNA molecules related to a particular gene. Most state of the art algorithms in this field like random forests or boosting ensembles can be seen as combination strategies for single threshold classifiers. The structure of these classifiers is beneficial in these high-dimensional settings as feature reduction is possible which also allows for a direct semantic and syntactic interpretation.

A single ray, the half-open interval representing one class, compares a single expression value of the profile with a threshold. In this work an alternative base classifier, the fold change classifier, is discussed. The classifier compares two expression values of the same sample. We analyze fold change classifiers as unweighted ensembles of type majority or unanimity vote. A sample compression bound for unweighted ensembles of fold change classifiers is also given.

1 Introduction

Modern diagnostic tools allow the categorization of tissues and other biological samples according to their gene expression profile. The classification of such data is based on high-dimensional vectors containing expression levels of thousands of RNA molecules. The classification models with the highest interpretability can directly be translated into natural language, such as

$$\text{IF } x_1 \geq 7 \text{ AND } x_5 \leq 4 \text{ OR } x_3 \geq 2 \text{ THEN } \ldots . \tag{1}$$

L. Lausser · H.A. Kestler (✉)

Research Group of Bioinformatics and Systems Biology, Institute of Neural Information Processing, University of Ulm, 89069 Ulm, Germany

e-mail: hans.kestler@uni-ulm.de

W. Gaul et al. (eds.), *German-Japanese Interchange of Data Analysis Results*,
Studies in Classification, Data Analysis, and Knowledge Organization,
DOI 10.1007/978-3-319-01264-3__17,
© Springer International Publishing Switzerland 2014

Typical models allowing such a syntactical interpretation are for example decision trees (Breiman et al. 1984) , decision lists (Rivest 1987) or unweighted ensembles of rays (Kestler et al. 2011). Modern algorithms utilizing these schemes are, e.g. random forests (Breiman 2001), boosting ensembles (Freund and Schapire 1995) or the set covering machine (Marchand and Shawe-Taylor 2002). The semantic interpretability of these models depends on the chosen kind of base classifier. A common choice is the single threshold classifier or ray. This base classifier compares a single measurement to a threshold (e.g. $x_1 \geq 7$). Although this base classifier is adequate in many applications, it is questionable if the single threshold classifier is the best choice for expression data. Its dependency on a fixed threshold makes a ray susceptible to global multiplicative or additive effects. Other algorithms use $x_i > x_j$ as a base classifier (Geman et al. 2004; Tan et al. 2005; Eddy et al. 2010). Bø and Jonassen (2002) utilize this pairwise criterion for feature selection and (Xu et al. 2008) utilize base classifiers of type $x_i > x_j > x_k$. In some way these procedures can be seen as extensions of the threshold approach.

In this work we discuss an alternative base classifier of type $\frac{x_i}{x_j} \geq t$. It will be called fold change classifier in the following. It is a natural extension of our conjunction of rays approach (Kestler et al. 2011). In fact, our motivation for studying this type of classification scenario originates from the often used paradigm of considering fold-expression changes between groups as a criterion of finding differentially regulated genes.

In the following we analyze the influence of fold change classifiers on unweighted ensembles of type majority vote or logical conjunction and give a sample compression bound for unweighted ensembles.

2   Methods

Throughout the rest of this paper the following notation will be used. An unlabeled sample will be given by $\mathbf{x} = (x_1, \ldots, x_n)^T$, $\mathbf{x} \in \mathbb{R}^n$. A class label y will be assumed to be binary $y \in \{0, 1\}$. The set of labeled training samples will be denoted $\mathbf{Z} = \{(\mathbf{x}_i, y_i)\}_{i=1}^m$. It is assumed that it was drawn from a fixed but unknown distribution $\mathscr{Z}$. For a single sample also $\mathbf{z} \in \mathbf{Z}$ will be used.

Single threshold classifiers (rays): The structure of a single threshold classifier $r_i^{(t)}(\mathbf{x})$ can be formalized by using the indicator function $\mathbb{I}_{[a]}$ evaluating to 1 if condition a is true and 0 otherwise.

$$r_i^{(t)}(\mathbf{x}) = \mathbb{I}_{[(x_i - t)d \geq 0]} \tag{2}$$

The decision criteria of ray $r_i^{(t)}(\mathbf{x})$ is restricted to a single feature value x_i which is compared to a threshold $t \in \mathbb{R}$. The direction $d \in \{-1, 1\}$ indicates if the relation of x_i and t should be $x_i \geq t$ (for $d = 1$) or $x_i \leq t$ (for $d = -1$).

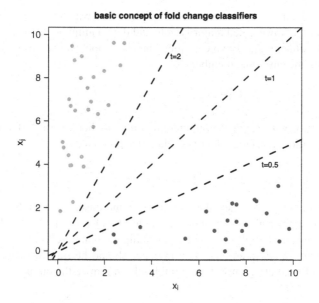

Fig. 1 Three fold change classifiers for features x_i and x_j ($t \in \{0.5, 1, 2\}$). Linear decision boundaries in a two dimensional subspace are generated

Fold change classifiers: A fold change classifier is given by following equation:

$$h_{ij}^{(t)}(\mathbf{x}) = \mathbb{I}_{\left[\frac{x_i}{x_j} \geq t\right]} = \mathbb{I}_{[x_i \geq t x_j]} \tag{3}$$

Here, the threshold $t \in \mathbb{R}_+$ is used to determine or set how large the relative difference between x_i and x_j should be. An illustration of a fold change classifier can be found in Fig. 1. The fold change classifier corresponds to a linear decision boundary in the two dimensional subspace of features i and j. For $t = 1$ it is equivalent to the bisecting line of the first quadrant.

Unweighted ensemble methods: In many cases the predictive power of a base classifier (either a single threshold classifier or a fold change classifier) is too low to get an accurate classification performance. In this case an ensemble of base classifiers $c_1(\mathbf{x}), \ldots, c_k(\mathbf{x})$ is trained and combined via some kind of fusion strategy, e.g. cart (Breiman et al. 1984), boosting (Freund and Schapire 1995), random forests (Breiman 2001), etc. The combined prediction is often of a higher accuracy than the prediction of an ensemble member. On the other hand, ensemble classifiers inherit some properties of their base classifiers. Among all possible fusion strategies the unweighted ensemble methods belong to those with the best interpretability. The methods treat all single predictions in the same manner. This implies that knowing all ensemble members is equivalent to knowing the consensus.

No additional parameters or weights are needed. Two well known members of this category are the unweighted majority vote and the unanimity vote.

An unweighted majority vote predicts the class label which was predicted most often among the ensemble members.

$$h(x) = \mathbb{I}_{[\sum c_i(x) > \sum (1 - c_i(x))]} \tag{4}$$

The unanimity vote (logical conjunction) only predicts class 1 if all ensemble members predict this class. Otherwise class 0 is predicted.

$$h(x) = \mathbb{I}_{[\bigwedge c_i(x)]} \tag{5}$$

For the unanimity vote a standard training of the ensemble members (e.g. minimizing the empirical risk) is not advisable (Kestler et al. 2011). The risk of misclassifying an example of class 1 is much higher than misclassifying an example of class 0. In this case an ensemble member should be trained to maximize the sensitivity of predicting class 1. It should only minimize the empirical risk on the remaining samples.

Data dependent classifiers: A subset of classifiers $\mathcal{H}(\mathbf{Z}) \subseteq \mathcal{H}$ of a hypothesis class $\mathcal{H}$ is called data dependent, if each $h \in \mathcal{H}(\mathbf{Z})$ can be reconstructed by utilizing a certain sequence of training examples, a so called compression set $\mathbf{Z}_c$, and some additional information given by a message string $\sigma \in \mathcal{S}$.

$$\rho : \mathcal{Z} \times \mathcal{S} \rightarrow \mathcal{H} \tag{6}$$

The mapping ρ is called a reconstruction function of $\mathcal{H}(\mathbf{Z})$. Although $\mathcal{H}(\mathbf{Z})$ is often much smaller than $\mathcal{H}$, a restriction to a set of data dependent classifiers is often preferable. It especially allows the application of sample compression bounds (N. Littlestone and M. Warmuth, 1986, Relating data compression and learnability, Unpublished manuscript), a kind of $(1 - \delta)$ confidence bound on the risk of a classification method. These bounds allow to calculate limits of the risk of a classification method, which will not be exceeded with a probability of $(1 - \delta)$. Other types of such bounds are, e.g., VC bounds (Vapnik 1998) or PAC-Bayes bounds (McAllester 1999). Sample compression bounds belong to the class of training set bounds, which means that they can directly be applied in the training phase of a classifier and do not depend on a separate test or validation set (test set bounds).

The idea of sample compression bounds is directly based on the concept of data dependence. By definition a data dependent classifier can be trained by only utilizing samples out of the compression set $\mathbf{Z}_c$. The remaining set of training samples $\mathbf{Z}_r = \mathbf{Z} \setminus \mathbf{Z}_c$ is unused in this context. The set $\mathbf{Z}_r$ can be seen as being independent from the training process of this particular classifier. The set $\mathbf{Z}_r$ can be used as a validation set for this classifier and can be applied as in a test set bound. This procedure can be repeated for each classifier in $\mathcal{H}(\mathbf{Z})$. An overall confidence of $(1 - \delta)$ for these bounds can be achieved by distributing the overall risk of failure δ among them. A more detailed introduction on sample compression bounds can be found in Langford (2005).

Theorem 1 (cf. Laviolette et al. 2006). *For a random sample* $\mathbf{Z} = (Z_1, \ldots, Z_m)$ *of iid examples drawn from* $\mathscr{Z}$, *and for every* $\delta \in (0, 1]$,

$$\Pr \left\{ \forall h \in \mathscr{H}(\mathbf{Z}): R(h) \leq \overline{\mathrm{Bin}} \left(R_{\mathrm{emp}}(h, \mathbf{Z}_r), \Pr_{\mathscr{Z}_c \times \mathscr{S}}(h)\delta \right) \right\} \geq 1 - \delta \qquad (7)$$

Loosely speaking the misclassification risk $R(h)$ of each data dependent classifier $h \in \mathscr{H}(\mathbf{Z})$ is bound by a term based on the binomial tail inversion.

$$\overline{\mathrm{Bin}} \left(\frac{k}{m}, \delta \right) = \sup \left\{ p \mid \mathrm{Bin} \left(\frac{k}{m}, p \right) \geq \delta \right\} \qquad (8)$$

Equation 8 is an $(1 - \delta)$ confidence bound on the risk p of a fixed classifier after receiving k errors in m independent tests. It can be seen as the inversion of the binomial tail, which denotes the probability of receiving an error rate k/m for a classifier with risk p.

$$\mathrm{Bin} \left(\frac{k}{m}, p \right) = \sum_{i=0}^{k} \binom{m}{i} p^i (1 - p)^{m-i} \qquad (9)$$

In Theorem 1 the confidence $(1 - \delta)$ is given to the set of bounds on each reconstructable classifiers and not to the bound of a particular one. As a consequence the error probability δ has to be distributed among all classifiers. This is done according to an arbitrary but fixed prior distribution $\Pr_{\mathscr{Z}_c \times \mathscr{S}}$. Kestler et al. (2011) showed sample compression bounds for majority votes and unanimity votes of data dependent rays. These compression bounds will now be extended to data dependent fold change classifiers and the union of both classifier types.

Data dependent fold change classifiers: A fold change classifier $h_{ij}^{(t)}(\mathbf{x}) = \mathbb{I}_{[x_i \geq t x_j]}$ will be called data dependent if there is a sample $\hat{\mathbf{z}} \in \mathbf{Z}$ with $\hat{x}_i / \hat{x}_j = t$

$$h_{ij}^{(t = \hat{x}_i / \hat{x}_i)}(\mathbf{x}) = \mathbb{I}_{\left[x_i \geq \frac{\hat{x}_i}{\hat{x}_j} x_j \right]}. \qquad (10)$$

The corresponding reconstruction function ρ^* only utilizes this single sample. Here the information string consists of an ordered tuple (i, j) of feature indices $i < j, i, j \in \{1, \ldots, n\}$ and a binary variable $d \in \{-1, 1\}$.

$$\rho^*(\{\hat{\mathbf{z}}\}, (i, j), d) = \begin{cases} h_{ij}^{(t = \hat{x}_i / \hat{x}_j)}(\mathbf{x}) & \text{if } d = 1 \\ h_{ji}^{(t = \hat{x}_j / \hat{x}_i)}(\mathbf{x}) & \text{else.} \end{cases} \qquad (11)$$

The value of d is used to determine the ordering of the two feature dimensions.

The reconstruction ρ^* can easily be extended to cover both data dependent rays and fold change classifier. The message string must therefore be allowed to contain a partially ordered tuple (i, j) of feature indices $i \leq j, i, j \in \{1, \ldots, n\}$. The combined reconstruction function will be called ρ^+ in the following.

$$\rho^+(\{\hat{z}\}, (i, j), d) = \begin{cases} \rho^*(\{\hat{z}\}, (i, j), d) & \text{if } i < j \\ \rho(\{\hat{z}\}, i, d) & \text{if } i = j \end{cases} \quad (12)$$

Data dependent majority votes or unanimity votes: As both ensemble types belong to the group of unweighted ensemble methods no additional information is needed to reconstruct the applied fusion scheme. The information therefore can be gained be concatenating the single compression sets and message strings. We are now ready to give a sample compression bound of majority votes or unanimity votes of data dependent rays or fold change features, which extends the formerly given compression bound in Kestler et al. (2011).

In the following data dependent majority votes or conjunctions will be denoted as $\mathcal{M}(\mathbf{Z})$ and $\mathscr{C}(\mathbf{Z})$. For the different base classifiers the notation will correspond to the notation given above (e.g. $\mathcal{M}^*(\mathbf{Z})$ and $\mathcal{M}^+(\mathbf{Z})$).

Theorem 2. *Let $\mathbf{Z} = (Z_1, \ldots, Z_m)$ denote a random sample of m examples drawn iid from $\mathscr{Z}$. Let $\mathscr{U}(\mathbf{Z})$ denote a set of unweighted ensembles of data dependent fold change classifiers based on the reconstruction functions described above. Then for every $\mathbf{Z}$ and for every $\delta \in (0, 1]$,*

$$\Pr\left\{\forall h \in \mathscr{U}(\mathbf{Z}) : R(h) \leq \overline{\text{Bin}}\left(R_{\text{emp}}(h, \mathbf{Z}_r), q(|h|)\delta\right)\right\} \geq 1 - \delta \quad (13)$$

where

$$q(|h|) = \frac{6m^{-|h|}2^{-|h|}}{\pi^2(|h| + 1)^2\binom{n(n-1)/2}{|h|}} \text{ for } \mathcal{M}^*(\mathbf{Z}) \text{ and} \quad (14)$$

$$q(|h|) = \frac{6m^{-|h|}2^{-|h|}}{\pi^2(|h| + 1)^2\binom{n(n+1)/2}{|h|}} \text{ for } \mathcal{M}^+(\mathbf{Z}). \quad (15)$$

Here $h \in \mathscr{U}(\mathbf{Z})$ denotes a single unweighted ensemble and $|h|$ the number of its ensemble members. The given bounds also hold for the sets of data dependent conjunctions $\mathscr{C}^*(\mathbf{Z})$, $\mathscr{C}^+(\mathbf{Z})$ by replacing the corresponding $q(k)$ of $\mathcal{M}^*(\mathbf{Z})$, $\mathcal{M}^+(\mathbf{Z})$ by $q(|h|) = q(|h|)|h|!$.

Proof. In order to derive the sample compression bounds from Theorem 1 we need to give reconstruction functions and prior distributions for the corresponding hypothesis classes. The reconstruction functions were given in Eqs. 11 and 12.

The prior distribution $P_{\mathscr{Z}_c \times \mathscr{S}}$ can be decomposed into the prior distribution over all compression sets $P_{\mathscr{Z}_c}$ and the distribution of all message strings for a fixed message string length $P_{\mathscr{S}}$. The prior $P_{\mathscr{Z}_c}$ is equivalent for both hypothesis classes. It is a distribution over compression sets of varying sizes $|h| \geq 0$. For a fixed $|h|$ the compression sets can be assumed to be uniform distributed. These distributions are merged to $P_{\mathscr{X}_c}$ by weighting the sizes by $6/(\pi(|h| + 1))^2$ (since $\sum_{|h|=1}^{\infty} |h|^{-2} = \pi^2/6 \leq 1$).

Once the compression set is fixed the string set $\mathscr{S}$ can be seen as uniformly distributed over the set of all allowed messages. For the set of fold change classifiers this corresponds to the set of pairs $(\mathbf{j}, \mathbf{d})$, where $\mathbf{j}$ is an ordered and repetition-free vector $\mathbf{j} \in \{(i, j) : 1 < i < j < n\}^{|h|}$ and $\mathbf{d}$ is a vector $\mathbf{d} \in \{-1, +1\}^{|h|}$. For the hypothesis class of combined classifiers the vectors $\mathbf{j}$ are allowed to be chosen out of $\mathbf{j} \in \{(i, j) : 1 < i \leq j < n\}^{|h|}$. If the ensemble is a conjunction for which the base classifiers were trained as proposed before we can further assume that the compression set is ordered though it may contain duplicates (Kestler et al. 2011).

$\square$

Examples for the sample compression bounds can be found in Table 2.

3 Simulation Experiments

The fold change classifiers were analyzed in 10×10 cross-validation experiments. They were tested in majority vote ensembles of 1 up to 1,000 ensemble members. The ensemble members were selected according to following objective $|\text{TPR}(h) - \text{FPR}(h)|$ which is the absolute difference of true positive rate TPR and false positive rate FPR. The classifiers with the highest values were chosen. We analysed ensembles of three different types of fold change classifiers. The first one denoted by $\text{maj}(t = i)$ utilizes fold change classifiers with parameter t fixed to $i \in \{0.1, 0.2, \dots, 2\}$. This category especially includes ensembles of type $h_{ij}^{(t=1)}$. The second one consists of data dependent fold change classifiers. Each fold change classifier is allowed to search for an optimal value of t within the set of reconstructable values. An ensemble of this type will be denoted by maj^*. The third type of ensemble utilizes a mix of data dependent fold change classifiers and single threshold classifiers. It will be denoted by maj^+.

The results of the ensembles are compared with random forests (Breiman 2001), support vector machines (Cortes and Vapnik 1995) and classification trees (Breiman et al. 1984). For the random forests (rf) the number of trees is varied from 1 up to 100. The support vector machines (svm) are used with linear svm(lin) and radial svm(rad) kernels. For both svms the cost parameter c is varied from 0.1 up to 10.0 by a step size of 0.1. For the classification trees (cart) the minimal number of samples per node are varied from 1 to 10. The experiments were performed on two microarray datasets.

The Shipp dataset (Shipp et al. 2002) contains 77 cancerous samples of a single B-cell lineage. The samples can be distinguished in 58 samples of diffuse large B-cell lymphoma (DLBCL) and 19 samples of GC B-cell lymphoma, follicular (FL).

Table 1 Results of the 10×10 cross-validation experiments for the Shipp and the West datasets. For all classifiers the average accuracy is reported. For those classifiers which are based on fold change features additionally the number of base classifiers is given

Shipp			West		
Classifier	Accuracy	# base cl.	Classifier	Accuracy	# base cl.
fcc($t = 1$)	95.45	1	fcc($t = 1$)	77.55	1
fcc($t = 0.3$)	99.74	1	fcc($t = 0.4$)	96.12	1
fcc*	98.05	1	fcc*	73.27	1
fcc$^+$	96.75	1	fcc$^+$	72.24	1
maj($t = 1$)	95.97	3	maj($t = 1$)	88.78	140
maj($t = 0.3$)	99.74	1	maj($t = 0.4$)	96.12	1
maj*	98.05	1	maj*	89.80	115
maj$^+$	96.75	1	maj$^+$	89.80	577
svm(lin)	96.75	–	svm(lin)	92.45	–
svm(rad)	90.65	–	svm(rad)	82.45	–
rf	89.61	–	rf	89.80	–
Cart	90.39	–	cart	87.76	–

Table 2 Values of the sample compression bound for different compression set sizes $|\mathcal{Z}_c|$

$|\mathcal{Z}_l|$	Error rate	Bound ($\delta = 0.05$)				
		$n = 50$	$n = 100$	$n = 150$	$n = 200$	$n = 1,000$
1	1.00	8.05	9.34	10.08	10.60	13.47
	5.00	15.76	17.29	18.16	18.76	21.96
	10.00	23.15	24.83	25.76	26.41	29.82
5	1.00	8.38	9.71	10.48	11.02	13.99
	5.00	16.38	17.97	18.86	19.48	22.78
	10.00	24.04	25.77	26.74	27.41	30.91
10	1.00	8.82	10.22	11.03	11.60	14.71
	5.00	17.23	18.88	19.82	20.47	23.91
	10.00	25.25	27.06	28.07	28.76	32.39

The West dataset (West et al. 2001) contains expression profiles from 49 breast cancer samples. These samples can be categorized according to their estrogen receptor status (25 ER+/24 ER−). The expression profiles are based on 7,129 features.

The results of the cross-validation experiments are summarized in Table 1. The table shows the results of the best single fold change classifiers (fcc, fcc*, fcc$^+$) which corresponds to the first ensemble members. Additionally the results of the best ensembles are shown. For the Shipp dataset, the single fold change classifiers outperformed rf, cart and svm(rad). The linear support vector machine outperformed fcc($t = 1$) and showed an equal accuracy to fcc$^+$. On this dataset the majority vote ensemble could only increase the accuracy for maj($t = 1$). For the West dataset only one fold-change-classifier (fcc($t = 0.4$)) could outperform the reference classifiers. The corresponding majority votes were able to increase the accuracies up to 17 %. They outperformed svm(rad) and cart.

4 Discussion and Conclusion

Many standard classification methods can be seen as combinations of simple base classifiers. A standard base classifier in this context is the single threshold classifier, e.g. $x_i \geq t$. Simply comparing a single feature value to a threshold t, it is one of the best interpretable classifiers. A drawback of this kind of classifier is its dependency on a specific threshold. It conceals implicit assumptions on the underlying data of being equally scaled and offset free which is often not the case. The single threshold classifier inherits its vulnerability to global multiplicative or additive influence to the combined classifiers.

The fold change classifier seems to be a valid alternative or even superior. Its relative decision criterion $\frac{x_i}{x_j} \geq t$ is also directly interpretable. Furthermore this classifier is invariant against scaling effects. Also the new classifiers can be trained in a data dependent way, which makes them suitable for analysis via sample compression bounds.

The cross-validation experiments show, that these base classifiers can be used to construct simple ensembles having a comparable accuracy to state of the art classifiers as random forests or support vector machines, but at the same time being of a much simpler structure. Moreover, the pairwise relation of genes can shed new light on possible relationships between involved pathways such as crosstalk. Furthermore, the simple comparison of expression values does not account for global differences in expression across genes as the increased performance for values of $t \neq 1$ suggests.

References

Bø T, Jonassen I (2002) New feature subset selection procedures for classification of expression profiles. Genome Biol 3(4):1–11

Breiman L (2001) Random forests. Mach Learn 45(1):5–32

Breiman L, Friedman JH, Olshen RA, Stone CJ (1984) Classification and regression trees. Wadsworth Publishing Company, Belmont

Cortes C, Vapnik V (1995) Support-vector networks. Mach Learn 20(3):273–297

Eddy JA, Sung J, Geman D, Price ND (2010) Relative expression analysis for molecular cancer diagnosis and prognosis. Technol Cancer Res Treat 9(2):149–159

Freund Y, Schapire RE (1995) A decision-theoretic generalization of on-line learning and an application to boosting. In: Vitányi P (ed) Proceedings of the second European conference on computational learning theory, Barcelona, vol 904. Springer, London, pp 23–37

Geman D, d'Avignon C, Naiman DQ, Winslow RL (2004) Classifying gene expression profiles from pairwise mRNA comparisons. Stat Appl Genet Mol Biol 3(9):1–16

Kestler HA, Lausser L, Lindner W, Palm G (2011) On the fusion of threshold classifiers for categorization and dimensionality reduction. Comput Stat 26:321–340

Langford J (2005) Tutorial on practical prediction theory for classification. J Mach Learn Res 6:273–306

Laviolette F, Marchand M, Shah M (2006) A PAC-Bayes approach to the set covering machine. In: Weiss Y, Schoelkopf B, Platt J (eds) NIPS, Vancouver, pp 731–738

Marchand M, Shawe-Taylor J (2002) The set covering machine. J Mach Learn Res 3(4–5):723–746

McAllester DA (1999) PAC-Bayesian model averaging. In: Ben-David S, Long P (eds) COLT '99: proceedings of the twelfth annual conference on computational learning theory, Santa Cruz. ACM, pp 164–170

Rivest RL (1987) Learning decision lists. Mach Learn 2(3):229–246

Shipp MA, Ross KN, Tamayo P, Weng AP, Kutok JL, Aguiar RCT, Gaasenbeek M, Angelo M, Reich M, Pinkus GS, Ray TS, Koval MA, Last KW, Norton A, Lister TA, Mesirov J, Neuberg DS, Lander ES, Aster JC, Golub TR (2002) Diffuse large b-cell lymphoma outcome prediction by gene-expression profiling and supervised machine learning. Nat Med 8(1):68–74

Tan AC, Naiman DQ, Xu L, Winslow RL, Geman D (2005) Simple decision rules for classifying human cancers from gene expression profiles. Bioinformatics 21:3896–3904

Vapnik V (1998) Statistical learning theory. Wiley, New York

West M, Blanchette C, Dressman H, Huang E, Ishida S, Spang R, Zuzan H, Olson JA, Marks JR, Nevins JR (2001) Predicting the clinical status of human breast cancer by using gene expression profiles. Proc Natl Acad Sci USA 98(20):11462–11467

Xu L, Tan A, Winslow R, Geman D (2008) Merging microarray data from separate breast cancer studies provides a robust prognostic test. BMC Bioinform 9(125):1–14

An Automatic Extraction of Academia-Industry Collaborative Research and Development Documents on the Web

Kei Kurakawa, Yuan Sun, Nagayoshi Yamashita, and Yasumasa Baba

Abstract This research focuses on an automatic extraction method of Japanese documents describing University-Industry (U-I) relations from the Web. The method proposed here consists of a preprocessing step for Japanese texts and a classification step with a SVM. The feature selection process is especially tuned up for U-I relations documents. A U-I document extraction experiment has been conducted and the features found to be relevant for this task are discussed.

1 Introduction

To make a policy of science and technology research and development, university-industry-government (U-I-G) relations are an important aspect to be investigated (Leydesdorff and Meyer 2003). It is one of the research targets to identify or clarify the U-I-G relations among organizations from the Web documents. In the clarification process, to get the exact documents relevant to U-I-G relations from the Web is the first requirement. The objective of this research is to extract automatically documents of U-I relations from the Web. We set a target into "press release articles" of organizations, and make a framework to automatically crawl them and decide which is of U-I relations.

K. Kurakawa (✉) · Y. Sun
National Institute of Informatics, Tokyo, 101-8430, Japan
e-mail: kurakawa@nii.ac.jp; yuan@nii.ac.jp

N. Yamashita
Japan Society for the Promotion of Science, Tokyo, 102-8472, Japan
e-mail: nagayoshi3@gmail.com

Y. Baba
The Institute of Statistical Mathematics, Tokyo, 190-8562, Japan
e-mail: baba@ism.ac.jp

W. Gaul et al. (eds.), *German-Japanese Interchange of Data Analysis Results*,
Studies in Classification, Data Analysis, and Knowledge Organization,
DOI 10.1007/978-3-319-01264-3__18,
© Springer International Publishing Switzerland 2014

In Sect. 2, we describe the framework which consists of a Web documents crawling process and a machine learning based document classification process.

2 Automatic Extraction Framework for Academia-Industry Collaborative R&D Documents on the Web

2.1 Crawling Web Documents and Extracting Texts

When we look for a document about U-I relations, especially in terms of collaborative research and development, there can be several sources of the form. We can find press release news sites of the organization, faculty introductory sites, laboratory home pages, researchers' own sites, such as blogs, and general scientific news sites of commercial sectors. Generally speaking, the sources of evidence of the fact that university and industry collaborate in research and development vary. However, we conclude that the press release site of organizations is promising source for automatic extraction and coverage. The reason is that in both university and industry articles of research and development results tend to be published in a news release, and the news articles are well-formed, so that it is better to analyze and classify these texts.

Figure 1 shows the automatic extracting process of documents on U-I relations from the Web. In the first step, we set seed URLs with popular crawling program such as wget[1] to crawl all press release news articles. Then we extract relevant texts from each article to classify whether it is documenting U-I relations or not. Extracted texts from the web documents are very noisy for content analysis. Even though we scrape the text from HTML tagged documents, irrelevant text, e.g. menu label text, header or footer of a page, or ads are still remaining. We have observed that the remaining irrelevant text fragments do not form complete sentences whereas the text containing U-I relations usually consists of two or three sequential, complete and well-formed sentences. For example, "the MIT researchers and scientists from MicroCHIPS Inc. reported that...", and off course in our real target of Japanese "東京大学とオムロン株式会社は、共同研究により、重なりや隠れに強く...." are picked up. In case of Japanese, in that sense, the necessary text fragments for detection are only those including punctuation marks which means fully formal sentence. Extracted text becomes fairly less noisy.

2.2 Classifier

We apply support vector machine (SVM) (Vapnik 1995) in order to decide whether a document is described U-I relations or not. SVM is a kernel-based algorithm that

---

[1]http://www.gnu.org/software/wget/

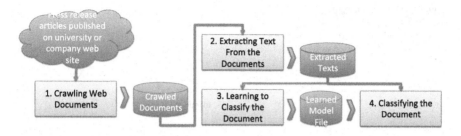

Fig. 1 Automatic extraction framework for U-I relations documents on the Web

have sparse solutions, so that predictions for new inputs depend only on the kernel function evaluated at a subset of the training data points. The determination of the model parameters corresponds to a convex optimization problem, and so any local solution is also a global optimum.

2.2.1 Support Vector Machine

A support vector machine is the two-class classifier simply using linear models of the form

$$y(\mathbf{x}) = \mathbf{w}^T \boldsymbol{\phi}(\mathbf{x}) + b. \tag{1}$$

where $\boldsymbol{\phi}(\mathbf{x})$ denotes a fixed feature-space transformation, and b is the bias parameter. The training data set comprises N input vectors $\mathbf{x}_1, \ldots, \mathbf{x}_N$, with corresponding target values $t_1, \ldots, t_N$ where $t_n \in \{-1, 1\}$, and new data points $\mathbf{x}$ are classified according to the sign of $y(\mathbf{x})$.

Assuming that the training data is linearly separable in feature space, so that by definition there exists at least one choice of the parameters $\mathbf{w}$ and b that satisfies $y(\mathbf{x}) > 0$ for points having $t_n = +1$ and $y(\mathbf{x}_n) < 0$ for points having $t_n = -1$, so that $t_n y(\mathbf{x}_n) > 0$ for all points.

To try to find the unique solution of $\mathbf{w}$ and b, SVM uses the concept of the *margin*, which is defined to be the smallest distance between the decision boundary and any of the samples, as illustrated in Fig. 2. The decision boundary is chosen to be the one for which the margin is maximized. The location of this boundary is determined by a subset of the data points (support vectors), which are indicated by the circle on the hyperplanes $y(\mathbf{x}) = 1$ and $y(\mathbf{x}) = -1$.

In this case, the optimization problem simply requires to maximize $\|\mathbf{w}\|^{-1}$, which is equivalent to minimize $\|\mathbf{w}\|^2$. Then we have to solve the optimization problem

$$\arg\min_{\mathbf{w},b} \frac{1}{2} \|\mathbf{w}\|^2. \tag{2}$$

Fig. 2 Maximize margin
between hyperplane
$y(\mathbf{x}) = 1$ and $y(\mathbf{x}) = 0$

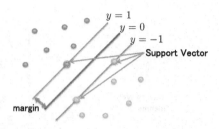

subject to the constraints, $t_n(\mathbf{w}^T\boldsymbol{\phi}(\mathbf{x}) + b) \geq 1, \quad n = 1,\ldots,N$ by means of the Lagrangian method.

In order to classify new data points using the trained data, we evaluate the sign of $y(\mathbf{x})$. This can be expressed by

$$y(\mathbf{x}) = \sum_{n=1}^{N} a_n t_n k(\mathbf{x}, \mathbf{x}_n) + b. \tag{3}$$

where the kernel function is defined by $k(\mathbf{x}, \mathbf{x}') = \boldsymbol{\phi}(\mathbf{x})^T\boldsymbol{\phi}(\mathbf{x}')$, and all $a_n > 0$ is Lagrange multipliers.

2.2.2 Feature Selection

Our problem is to classify web documents of U-I relations by means of SVM, so that the text document should be represented by a vector $\mathbf{x}_n$. The mapping from text document to a vector is known as feature selection, and several methods are proposed, i.e. term selection based on document frequency, information gain, mutual information, a χ^2-test, and term strength (Yang and Pedersen 1997). In our approach, we adopt tf-idf (term frequency – inverse document frequency), which is one of the most commonly used term weighting schemes in today's information retrieval systems (Aizawa 2003).

tf-idf is formally defined as follows.

$$\text{tf-idf}(t, d, D) = \text{tf}(t, d) \times \text{idf}(t, D). \tag{4}$$

t, d, D respectively denote a term, a document, and a corpus (set) of documents. $\text{tf}(t, d)$ is the number of occurrences of a term in a document d. $\text{idf}(t, D)$ is the inverse document frequency.

The feature is expressed by

$$\mathbf{x}_d = (x_{t_1,d}, x_{t_2,d}, \cdots, x_{t_M,d}), \quad x_{t,d} = \text{tf-idf}(t, d, D) \times b_{t,d}. \tag{5}$$

$b_{t,d}$ means the boolean existence of the terms in d. The term can be a term in a document, type of POS (part-of-speech) of morpheme, or analytical output of external tools in our experiment. The feature selection is the term selection, where the terms refers to the elements of $\mathbf{x}_d$. The feature $\mathbf{x}_d$ is calculated for each document, and the size of which depends on the term selection.

3 Features and SVM Parameters

We seek several features for SVM input for both learning and classifying. In this experiment, tf-idf is the base element of input vectors $\mathbf{x}_n$, so that the variations here are depending on term selections. Table 1 shows feature and SVM parameter selections for each classification test. For dividing a sequential Japanese text into morphemes, we use Mecab, a Japanese morphological analyzer (Kudo et al. 2004).

In the first group of features, three types of Bag of Words ((1). BoW) features are examined. The first one is that full output of Mecab is used to construct the BoW feature, and each word tf-idf consists of the feature vector $\mathbf{x}_n$. In the second ((2). BoW(N)), only the noun is chosen, and in the third ((3). BoW(N-3)), the word is restricted to proper-noun, general-noun, and Sahen-noun (verb formed by adding "する" ([suru], do) to the noun).

In the fourth ((4). K(14)), we prepare 14 keywords related to U-I relations and calculate tf-idf for all documents. Hence, the other words are ignored. The keywords are "研究" ([kennkyu], research), "開発" ([kaihatsu], development), "実験" ([jikken], experiment), "成功" ([seikou], success), "発見" ([hakken], discover), "開始" ([kaisi], start), "受賞" ([jushou], award), "表彰" ([hyoushou], honor), "共同" ([kyoudou], collaboration), "協同" ([kyoudou], cooperation), "協力" ([kyouryoku], join forces), "産学" ([sangaku], UI relationship), "産官学" ([sankangaku], UIG (University-Industry-Government) relations), and "連携" ([renkei], coordination). In the fifth ((5). K(18)), some keywords added to the preliminary 14, i. e. "受託" ([jutaku], entrusted with), "委託" ([itaku], consignment), "締結" ([teiketsu], conclusion), and "研究員" ([kennkyuin], researcher). In the sixth ((6). K(18) + NM), keywords and POS (Part of Speech) of the next morpheme in a sequential text are checked, in that grammatic connections of those keywords are restricted to verb, auxiliary verb, and Sahen-noun. In the seventh ((7). Corp.), cooperation marks are caught for the evidence of company name. Those are "株式会社"([kabushikigaisha], Incooperated), （株）(an unicode character as U+3231), （株), or (株). This corresponds to Ltd. or Inc. in English. In the eighth ((8). Univ.), university name is checked, i.e. "大学"([daigaku], university), or "大"([dai], a shorten representation of university). In the ninth ((9). C. + U.), it is checked whether both cooperation mark and university name are being in a sentence. In the tenth ((10). ORG), we checked the existing of an organization by means of Cabocha's Japanese named entity extraction function (Kudo and Matsumoto 2002).

Table 1 Feature selection and SVM kernel functions

| | TF-IDF feature element | | | | | | | | | | |
Test ID	(1) BoW	(2) BoW(N)	(3) BoW(N-3)	(4) K(14)	(5) K(18)	(6) K(18)+NM	(7) Corp.	(8) Univ.	(9) C.+U.	(10) ORG	Kernel
1-1	✓										Linear
1-2		✓									Linear
1-3			✓								Linear
2-1				✓							Linear
2-2				✓							Polynomial
2-3				✓							RBF
3-1					✓						Linear
3-2					✓						Polynomial
3-3					✓						RBF
4-1						✓					Linear
4-2						✓					Polynomial
4-3						✓					RBF
5-1						✓				✓	Linear
5-2						✓				✓	Polynomial
5-3						✓				✓	RBF
6-1						✓	✓	✓		✓	Linear
6-2						✓	✓	✓		✓	Polynomial
6-3						✓	✓	✓		✓	RBF
7-1						✓	✓	✓	✓		Linear
7-2						✓	✓	✓	✓		Polynomial
7-3						✓	✓	✓	✓		RBF
8-1						✓	✓	✓	✓	✓	Linear
8-2						✓	✓	✓	✓	✓	Polynomial
8-3						✓	✓	✓	✓	✓	RBF

Refer to the body of the section to know what abbreviations mean

Since the support vector machine is driven by a kernel function, we examine the linear kernel, polynomial kernel and RBF (Radial Basis Function) kernel here. We use SVM^{light} which is Joachims' implementation.[2]

4 Experiment to Classify

4.1 Data Set and Results

We prepared real web documents for the experimental data set used. Press release documents of several kinds of organizations are crawled and manually tested whether it is U-I relations or not. Then we selected the same amount of articles in both positives and negatives. The numbers are shown in the Table 2.

---

[2]http://svmlight.joachims.org/

Table 2 Data set for experiment

Organization	Crawled articles		Articles for experiment	
	Positive article	Negative article	Positive article	Negative article
Tohoku Univ.	44	499	44	44
The Univ. of Tokyo	106	848	106	106
Kyoto Univ.	40	329	40	40
Tokyo Inst. of Tech.	37	343	37	37
Hitachi Corp.	103	450	103	103
Total	**330**	**2,469**	**330**	**330**

Table 3 Classification results (average points in tenfold cross validation)

Test ID	Accuracy	Precision	Recall	F-measure
1-1	61.21	64.04	42.12	47.28
1-2	60.61	63.75	40.00	45.54
1-3	61.52	67.44	40.00	46.72
2-1	67.58	72.02	61.52	63.70
2-2	58.03	69.76	23.33	34.45
2-3	66.51	62.53	86.37	71.89
3-1	68.18	72.02	63.33	64.78
3-2	57.88	69.00	23.03	34.08
3-3	66.67	62.22	88.18	72.43
4-1	70.61	74.66	63.64	67.40
4-2	–	–	–	–
4-3	70.76	65.49	90.30	75.66
5-1	70.61	74.61	63.64	67.31
5-2	–	–	–	–
5-3	70.76	65.49	90.30	75.66
6-1	–	–	–	–
6-2	–	–	–	–
6-3	70.15	64.64	93.64	76.09
7-1	78.79	85.01	71.52	76.99
7-2	–	–	–	–
7-3	72.27	66.07	**94.85**	**77.61**
8-1	**78.94**	**85.03**	71.82	77.16
8-2	–	–	–	–
8-3	71.82	65.73	**94.85**	77.35

–Not calculated because of precision zero or learning optimization fault

For each combination of feature elements in Table 1, we then conducted tenfold cross validation for the above data set. Classification test results are shown in Table 3 from the viewpoint of accuracy, precision, recall, and f-measure. In the test ID 1-1, 1-2, 1-3, feature elements consist of bag of words (BoW) which count over 15,800, 13,000, and 12,000 words, respectively. The f-measures are worse than the other features with the same linear kernel function. The bags of words used in the first three tests seem to be unsuitable for learning.

In the test ID from 2-1 to 8-3, feature element size is about 14–33. They seem to be effectively caused by learning, except for some tests resulting in some learning optimization fault. Accuracy and f-measure are gradually improved while feature elements are additionally complex. The test ID 7-3 has the best f-measure, and it's recall rate is high too. But, when comparing the accuracy, test ID 8-1 is better than ID 7-3. Test ID 7-3 is totally better than others.

4.2 Discussion

In comparing with BoW features and keyword features, BoW features were unsuitable for learning, while keyword features produced good results. Even though BoW is known for useful general elements for text categorization, this time BoW features were unsuitable for learning. The reason they failed can be that the training data size was too small. If we have enough training data, it becomes larger than the feature vector size. This means training data size surpasses the number of basis function of SVM, so that learning could be done without over-fitting. In case of keyword features, the effects of learning were observed. This might be because keyword features construct small dimension of the feature vector comparing to the training data size.

To reduce the number of feature elements to the order 10^1, we have to choose the most sensitive terms as feature elements. In our study, the 18 keywords of U-I relations which we chose were effective for learning. The test ID 4-* shows the effectiveness of co-occurence of keywords and POS of the following morpheme. The score of test ID 4-* considering the 18 keywords and POS was 3 points higher than that of test ID 3-* considering only the 18 keywords.

In comparing test ID 4-* to test ID 5-*, the scores of them were equivalent. This means organization tag count as a result of named entity extraction is meaningless. In the press release articles, organization names may intend to be written in any genre of articles irrelevant to U-I relations.

Test ID 6-* and 7-* are based on the occurrence of university and company symbols. Especially in ID 7-3, recall and f-measure are the highest. This means the occurrence of the two symbols in a sentence is sensitive to U-I relations.

Comparing test ID 7-* and 8-*, what we can find is the effectiveness of the organization tag. But, effectiveness is not clear as mentioned in a previous sentence.

Finally, we want to mention the kernel function type. We understand it strongly depends on scores. Relations among kernel function type, feature elements, and training data size may reveal mechanisms of learning for this problem. This point remains for next research topic.

5   Conclusion

To extract automatically documents of U-I relations from the Web, we designed an automatic extraction framework that consists of a crawling process for press release articles of organizations and a machine learning process to detect if each article is documenting U-I relations or not. A support vector machine (SVM) is adopted for the learning process.

We designed feature vectors for SVM and conducted machine learning experiments to see the effect of several combinations of feature vector elements and kernel function types of SVM. The result is that U-I relations keywords, university and company symbols in a sentence are effective features for the detection of U-I relations. In the experiment of detecting university-industry cooperation we got an accuracy 78.94, f-measure 77.16 for classifying U-I relations documents on the Web. In the future work, we will analyze how kernel function type and the parameters effect on the scores.

References

Aizawa A (2003) An information-theoretic perspective of tf-idf measures. Inf Process Manag 39(1):45–65. doi:10.1016/S0306-4573(02)00021-3

Kudo T, Matsumoto Y (2002) Japanese dependency analysis using cascaded chunking. In: Roth D, van den Bosch A (eds) CoNLL-2002, Taipei, pp 63–69

Kudo T, Yamamoto K, Matsumoto Y (2004) Applying conditional random fields to Japanese morphological analysis. In: 2004 conference on empirical methods in natural language processing (EMNLP-2004), Barcelona, pp 230–237. http://mecab.googlecode.com/svn/trunk/mecab/doc/index.html

Leydesdorff L, Meyer M (2003) The triple helix of university-industry-government relations. Scientometrics 58(2):191–203

Vapnik VN (1995) The nature of statistical learning theory. Springer, New York

Yang Y, Pedersen JO (1997) A comparative study on feature selection in text categorization. In: Fisher DH (ed) Proceedings of ICML-97, 14th international conference on machine learning, Nashville. Morgan Kaufmann Publishers, San Francisco, pp 412–420

The Utility of Smallest Space Analysis for the Cross-National Survey Data Analysis: The Structure of Religiosity

Kazufumi Manabe

Abstract The purpose of this paper is to illustrate the utility of Smallest Space Analysis (SSA) developed by Louis Guttman using the examples of National Religion Surveys conducted in Japan (2007) and Germany (2008) by the research team organized by the author.

By conducting a data analysis of these two surveys, we try to examine the similarities and differences in religiosity between people in Japan and Germany. Smallest Space Analysis (SSA) demonstrates its usefulness in exploring the characteristics of religiosity in different countries from a comparative perspective.

1 Introduction

Since the emergence of secularization theory, religious pluralism theory, and religious market theory, there has been a revitalization of discussions on religiosity in the Western sociology of religion. A review of past studies reveals that among the studies of religion conducted in Western sociology, few have been conducted from the perspectives of cross-national comparison with Asian countries. Recognizing this problem, we have focused on the comparison of Japan and Germany to improve this research situation (Manabe 2012).

Based on the previous researches, we divided religiosity into two components: (1) religious practice, participation and behavior, and (2) religious faith, belief and feeling, and transformed these components into questionnaire items. Then, we conducted a nationwide survey in Japan using the questionnaire we constructed, and conducted a nationwide survey using essentially the same questionnaire in Germany (Manabe 2008, 2010).

K. Manabe (✉)
School of Cultural and Creative Studies, Aoyama Gakuin University, Tokyo, Japan
e-mail: kazufumi.manabe@nifty.com

W. Gaul et al. (eds.), *German-Japanese Interchange of Data Analysis Results*,
Studies in Classification, Data Analysis, and Knowledge Organization,
DOI 10.1007/978-3-319-01264-3_19,
© Springer International Publishing Switzerland 2014

This paper presents the results of the data analyses of these two surveys, from a comparative perspective. The comparative perspective is the perspective from which we examine the "similarities" and "differences" in religiosity between people in Japan and Germany (Jagodzinski and Manabe 2009; Manabe 2011).

The basic approach to data analysis used in this paper, figuratively speaking, is to begin by looking at the forest rather than the trees. The data analysis of questionnaire surveys generally begins with the identification of the overall structure of data, and this is followed by efforts to intensify the analysis by focusing on specific aspects of the data. I refer to the former structural aspects as "looking at the forest," and the latter more specific aspects as "looking at the trees." What kinds of techniques, then, can be used to "look at the forest"? Smallest Space Analysis (SSA) developed by Louis Guttman is a very useful tool in examining the "similarities" and "differences" of the "forests" when conducting a cross-national comparative survey (Levy 1994; Manabe 2001; Shye 1978).

As a type of multidimensional scaling, SSA is a method of expressing the relationship between n question items shown in a correlation matrix by the size of the distance between n points in an m-dimensional ($m < n$) space. The higher the correlation, the smaller the distance, and the lower the correlation, the greater the distance. Usually a 2-dimensional (plane) or 3-dimensional (cube) space is used to visually depict the relationship between question items. This shows that SSA is an appropriate method of visually depicting the overall structure and relationships among question items (For the technical and mathematical aspects, see Borg and Shye 1995).

2 Survey Outline

National surveys were conducted in both Japan and Germany using essentially the same questionnaire. Overviews of the surveys follow.

2.1 Japanese Survey

1. Respondents: Male and female adults aged 20 and older, nationwide.
2. Sampling procedure: Central Research Services was entrusted to conduct the sampling and surveying. The sample was obtained using a two-stage stratified random sampling of men and women aged 20 and older from the Basic Resident Registry as of March 31, 2006. The country was divided into 12 regions. These were divided by size into 16 large cities, other cities, and towns and villages, and surveys were conducted in 25 locations in the 16 large cities, 63 locations in the other cities, and 12 locations in the towns and villages, for a total of 100 locations. Eighteen people were surveyed from each of these 100 locations, yielding a total of 1,800 respondents.

3. Survey method: Questionnaire leave-and-pick-up method
4. Survey period: March 2007.
5. Valid response rate: 882/1,800 (49.2 %)

2.2 German Survey

1. Respondents: Male and female German-speaking adults aged 18 and older, nationwide.
2. Sampling procedure: German research company Marplan was entrusted to conduct the sampling and surveying. Survey respondents were selected using the Kish Method, from randomly selected households chosen using the random walk method, based on the ADM sampling system developed by the German Market and Public Opinion Research Institute (Arbeitskreis Deutscher Markt und Meinungsforschunginstitute, ADM) in the 1970s. Six residences were extracted from each of 129 ADM sample points nationwide (105 in West Germany, 24 in East Germany). Next, after subtracting from these 774 households those 42 where the survey could not be conducted due to vacancy or other reasons, survey respondents were selected using the Kish Method from the remaining 732 households. However, another 176 households were further excluded because "No one could be contacted after three visits" or because "The residents were unwilling to participate." Respondents were extracted from the remaining 556 households, and valid responses were received from 515 individuals (subtracting 41 individuals who could not respond to the survey due to illness or unwillingness to participate).
3. Survey method: Personal interview method using a questionnaire
4. Survey period: February to March 2008.
5. Valid response rate: 515/732 (70.4 %)

3 Results of Data Analysis

3.1 Religious Behaviors

For both Japan and Germany, I created a correlation matrix showing the relationships between the question items about "religious behavior" (10 items in Japan and 8 in Germany) and conducted a Smallest Space Analysis (SSA) using the Hebrew University Data Analysis Package (HUDAP), a computer software package for analyzing data. The results produced two 2D SSA maps (spatial partitions), as shown in Figs. 1 and 2. The original format of these SSA maps (computer output) had the number of each variable (question item) marking its position on the 2D space (Euclidean space). However, there were also three concentric circles drawn

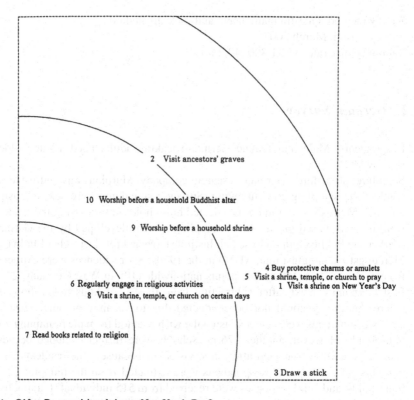

1   Q10   Do you visit a shrine on New Year's Day?
2   Q12a Visit ancestors' graves during the obon and higan seasons.
3   Q12b Draw a stick with a number on it to learn my fortune.
4   Q12c Buy protecitive charms or amulets (for traffic safety, passing entrance exams,etc.)
5   Q12d I visit a shrine, temple, or church to pray for such things as business prosperity and success in
           passing the entrance exams.
6   Q12e I am regularly involved in relgious activities, such as worship and devotions.
7   Q12f I read books related to religion, such as the Bible or sacred scriptures.
8   Q12g I visit a shrine, temple, or church on certain days.
9   Q12h I worship before a household shrine.
10 Q12i I worship before a household Buddhist altar.

Fig. 1 Smallest Space Analysis of religious behaviors (Japan)

in the space. These are the result of my efforts to apply meaning to (interpret) the spatial partition of the question items. How can these results be interpreted? The SSA maps for both Japan and Germany consist of spatial partitions in which the question items are plotted in three concentric circles centered around the item "I read books related to religion, such as the Bible," based on the given question item's relationship to the center item (expressed by the correlation coefficient), extending from the concentric circle closest to the center (containing items with larger correlation coefficients) to the concentric circle furthest away (containing items with smaller correlation coefficients). The reason circles are used rather than ovals for the space partitioning is because a circle expresses equidistance from the

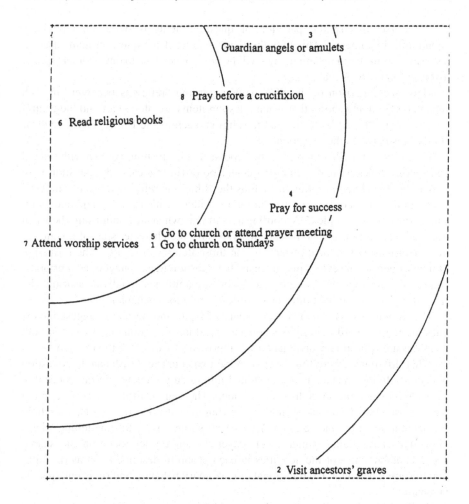

1 Q9   Do you attend worship services on important church holidays?

2 Q11-1 Visit ancestor's or relatives' graves on memorial days.

3 Q11-2 I have guardian angels or amulets with me to keep from everyday harm.

4 Q11-3 I go to church and pray for such tings as success at work and success in passing my exams.

5 Q11-4 I go to church or attend prayer meetings.

6 Q11-5 I read religious books, such as the Bible.

7 Q11-6 I go to church on Sundays and holidays.

8 Q11-7 I pray before a crucifixion (a statue of Christ) or other figure at home.

Fig. 2 Smallest Space Analysis of religious behaviors (Germany)

center (in this case, from the question item "I read books related to religion, such as the Bible"), thus reflecting that items have the same approximate size of correlation coefficients.

Of course, this spatial partition of question items is based on Guttman's "contiguity hypothesis." This hypothesis means that if the question items used in a survey are similar in meaning, they will be positioned close to one another (spatial distance) on the SSA map.

Thus, a comparison of the SSA maps regarding "religious behavior" in both Japan and Germany shows that when question items are plotted around the central question item "I read books related to religion, such as the Bible," there are clear similarities between the two countries.

The shape of the SSA map described above, that is, the figure drawn with several concentric circles around a common center point (in this case, the question item "I read books related to religion, such as the Bible") is referred to as a "simplex." Guttman calls the correlation structure among items which can be explained with simple rank order "simplex." According to Guttman, when the relationships between items addressed in this kind of data analysis have the characteristics of a simplex, the performance of a "Scale Analysis" on those items will result in the formation of a unidimensional scale. This suggests that question items categorized a priori as "question items regarding religious behavior" are unidimensionally measuring "the same thing" (behavior that must be regarded as religious behavior).

The similarities evident on the SSA maps of Japan and Germany suggest that the question items regarding religious behavior used in this nationwide survey in both countries are appropriate, and that they comprise a so-called "Guttman scale."

What differences were found between the two countries? In Japan, ten question items on religious behaviors are separated into three groups within the concentric circles into which the space has been divided. These are (a) Bible /sacred scripture reading, worship, devotions, shrine/temple/church visits, (b) grave visits, household shrine/altar worship, and (c) New Year's Day shrine visits, fortune-telling sticks, protective charms and amulets, and prayer. I then further focus on the shared social characteristics of the activities in each group to describe them as (a) faith-manifestation behavior, (b) traditional or customary behavior and (c) event-specific behavior.

While this reflects the pattern that emerged in Japan, the pattern that emerged in Germany is as follows: (a) Bible reading, church/prayer meetings, Sunday and holiday worship services, prayer before a crucifixion (a statue of Christ), (b) prayer for success, guardian angels or amulets, and (c) grave visits.

A comparison of the results obtained from Japan and Germany shows that there are similarities between the two countries insofar as the question items on religious behavior are able to be divided into three groups using three concentric circles, based on the size of the correlation coefficient between each item and the center item "I read books related to religion, such as the Bible".

However, there are important differences between the two countries with regard to the meanings of the questions that are included in each of the three groups. In Japan, the concentric circles around the center point start from "faith-manifestation behaviors" to "traditional behaviors," to "event-specific behaviors." In Germany however, there are doubts regarding the suitability of these labels, and the following labels would be proposed instead: from "faith-manifestation behaviors" to "instrumental behaviors," to "memorialism behaviors."

3.2 Religious Faith, Beliefs and Feelings

A comparison of the two countries' SSA maps (Figs. 3 and 4) shows that while the spatial partition of several question items do not indicate complete agreement in the religiosity of the Japanese and Germans (these will be discussed later), the major overall patterns are quite similar.

As has been done in previous analyses, I have tried to read the SSA map spatial partition based on the size of the correlation coefficients between the various question items and the center item, "I have religious faith." Again, several concentric circles have been drawn around that center point on each SSA map, with the innermost circle containing question items with the largest correlation coefficients with "I have religious faith" and the outermost circle containing question items with the smallest correlation coefficients with that item. Let's look at the similarities that can be seen in the two countries' maps.

1. Responses to most of the question items plotted in the second concentric circle in Germany and in the second and third concentric circles in Japan (counting outward from the center) are similar in both countries.
2. Question items on "religious views of nature," "memorialism," "the importance of the here and now," "art=religion," and "sixth sense" are plotted in the fourth and fifth circles in Japan, and in the third circle in Germany.
3. On both countries' SSA maps, the two items "All gods are the same" and "Happiness in this life is more important than salvation in the next life" were plotted in the outermost concentric circle.

This suggests that the overall structures of religiosity in Japan and Germany are similar, and should not be viewed as completely different. This is not to say that there are no differences, however, and those that do exist are outlined below.

1. The question item about "a religious mind" which was thought as a unique question in Japan was plotted all on its own in the second concentric circle on the Japanese map, quite far away from the question items contained in the third circle, but it was plotted in the second concentric circle on the German map, along with other question items in that sector.
2. Of the question items plotted in the second concentric circle on the German map:

   (a) "I feel a strong spiritual connection with my deceased relatives" is located more centrally in Japan, but more peripherally in Germany.
   (b) The three question items "I cannot litter near or contaminate a church," "If you are irreverent to God, you will be punished," and "If you do something wrong, even if no one else sees it, you will be punished" were plotted fairly

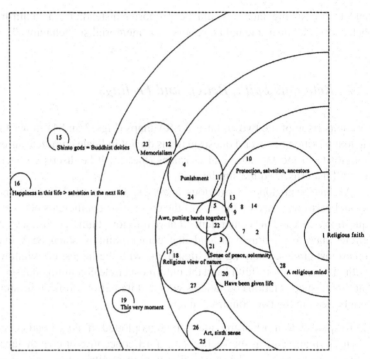

1   Q7   I have religious faith.
2   Q20a   I feel a strong connection to my ancestors.
3   Q20b   I feel a sence of peace when I visit a shrine, temple, or church.
4   Q20c   I cannot litter near, or contaminate a shrine, temple, or church.
5   Q20d   When I visit a shrine, temple, or church, I instinctively want to put my hands together.
6   Q20e   When I'm having problems, I call out in my heart to a god or deity.
7   Q20f   I'm grateful to the gods for my safe and peaceful daily life.
8   Q20g   When I pray to a Shinto or Buddhist deity,I feel like they somehow amswer my prayer.
9   Q20h   The souls of our ancestors are living on somewhere and are always watching out for us.
10 Q21a   Failure to hold a memorial service for one's ancestors is evidence of a lack of belief.
11 Q21b   If you are irreverent to gods or deities you will be punished.
12 Q21c   The memories of family members who have died are precious.
13 Q21d   Being saved by a god or deity means that things will go well for you in this life.
14 Q21e   Good behavior in this life will be rewarded in the next life.
15 Q21f   Shinto gods and Buddhist deities are all the same thing.
16 Q21g   Happiness in this life is more important than salvation in the next life.
17 Q21h   When looking at a large, old tree, a kind of feeling of divinity comes over me.
18 Q21i   At sunrise or sunset, or in the light of the moon, I experience a sense of solemnity.
19 Q21j   I think that this very moment here and now is an important time.
20 Q21k   I feel that I have been given life by a great power that I cannot see.
21 Q21l   When I hear a worship song or gospel music, or the singing of sutras or sacred songs, I :
            a sense of peace and solemnity.
22 Q21m   Gods and deities inspire awe.
23 Q21n   When I worship at a Buddhist altar or visit family graves, I think more about my deceas
            parents and grandparents than about my ancestors.
24 Q21o   If you do something wrong, even if no one else sees it, you will be punished.
25 Q21p   Sometimes I feel as though I have a sixth sense in which I am informed of what is going
            happen in the future.
26 Q21q   An excellent piece of artwork can convey a sense of something religious.
27 Q21r   I think that souls inhabit everything, such as mountains, rivers, grass, and trees.
28 Q24   A religious mind is importamt

Fig. 3 Smallest Space Analysis of religious beliefs, feelings and attitudes (Japan)

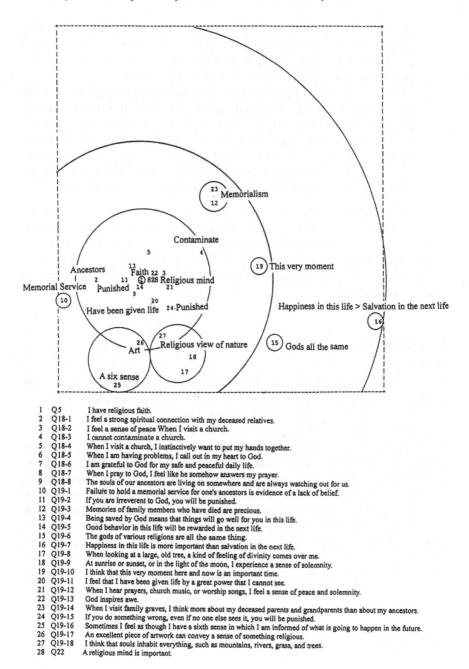

1	Q5	I have religious faith.
2	Q18-1	I feel a strong spiritual connection with my deceased relatives.
3	Q18-2	I feel a sense of peace When I visit a church.
4	Q18-3	I cannot contaminate a church.
5	Q18-4	When I visit a church, I instinctively want to put my hands together.
6	Q18-5	When I am having problems, I call out in my heart to God.
7	Q18-6	I am grateful to God for my safe and peaceful daily life.
8	Q18-7	When I pray to God, I feel like he somehow answers my prayer.
9	Q18-8	The souls of our ancestors are living on somewhere and are always watching out for us.
10	Q19-1	Failure to hold a memorial service for one's ancestors is evidence of a lack of belief.
11	Q19-2	If you are irreverent to God, you will be punished.
12	Q19-3	Memories of family members who have died are precious.
13	Q19-4	Being saved by God means that things will go well for you in this life.
14	Q19-5	Good behavior in this life will be rewarded in the next life.
15	Q19-6	The gods of various religions are all the same thing.
16	Q19-7	Happiness in this life is more important than salvation in the next life.
17	Q19-8	When looking at a large, old tree, a kind of feeling of divinity comes over me.
18	Q19-9	At sunrise or sunset, or in the light of the moon, I experience a sense of solemnity.
19	Q19-10	I think that this very moment here and now is an important time.
20	Q19-11	I feel that I have been given life by a great power that I cannot see.
21	Q19-12	When I hear prayers, church music, or worship songs, I feel a sense of peace and solemnity.
22	Q19-13	God inspires awe.
23	Q19-14	When I visit family graves, I think more about my deceased parents and grandparents than about my ancestors.
24	Q19-15	If you do something wrong, even if no one else sees it, you will be punished.
25	Q19-16	Sometimes I feel as though I have a sixth sense in which I am informed of what is going to happen in the future.
26	Q19-17	An excellent piece of artwork can convey a sense of something religious.
27	Q19-18	I think that souls inhabit everything, such as mountains, rivers, grass, and trees.
28	Q22	A religious mind is important.

Fig. 4 Smallest Space Analysis of religious beliefs, feelings and attitudes (Germany)

close to one another on the Japanese map, all within a single concentric circle, forming a kind of subgroup. On the German map, however, these three are distributed in scattered locations throughout the circles. Clearly there is little relationship of meaning between these three items.

(c) On the German map, the question items "When I am having problems, I call out in my heart to God," "I am grateful to God for my safe and peaceful daily life," and "When I pray to God, I feel like he somehow answers my prayer," were all plotted in the exact same location. As a result, we had a technical problem since the item numbers marking each item were all overlapping, leaving the last item number as the only one visible. This means that the relationship of meaning between these three items is extremely strong.

3. The item "Failure to hold a memorial service for one's ancestors is evidence of a lack of belief" was plotted in a concentric circle near the center point of "I hold religious beliefs" on the Japanese map, but was plotted in a circle located quite far from the center point on the German map. This reflects a substantive difference in meaning between "ancestor commemoration" in Japan and "marking memorial days" in Germany.

4 Conclusions

We have illustrated the utility of SSA for the cross-national comparison of religiosity in Japan and Germany. Then, what are the issues that will need to be addressed in the future? When reporting on the findings above, I tried to develop several hypothetical arguments. Many of these arguments addressed not only substantive problems related to religiosity in both countries, but also methodological problems, such as the wording of the question items used in the surveys. When preparing the German translation of the national questionnaire conducted in Japan, efforts were made to maintain the greatest degree of "functional equivalence" possible on the question items. Still, some problems undeniably remained. The data analysis of the national surveys conducted in both countries revealed that the substantive meaning of question items used in Japan and Germany were in some cases quite different. Careful investigation of the equivalence of meaning of the terms used in international comparative surveys is the greatest research challenge remaining to be addressed in the future.

References

Borg I, Shye S (1995) Facet theory: form and content. Sage, Thousand Oaks
Levy S (ed) (1994) Louis Guttman on theory and methodology: selected writings. Dartmouth, Aldershot/Brookfield
Jagodzinski W, Manabe K (2009) On the similarity in different cultures. In: Haller M, Jowell R, Smith TW (eds) The international social survey programme 1984–2009: charting the globe. Routledge, London/New York, pp 313–336

Manabe K (2001) Facet theory and studies of Japanese society: from a comparative perspective. Bier'sche Verlangsanstalt, Bonn

Manabe K (2008) The structure of religiosity: data analysis of a national survey on values and religiosity. In: Political science at Keio University: essays in celebration of the 150th anniversary of the founding of Keio University. Keio University Press, Tokyo, pp 283–326

Manabe K (2010) The structure of religiosity: a cross-national comparison of Japan and Germany. J Law Politics Sociol 83(2):448–494

Manabe K (2011) Cross-national comparison of the dimension and structure of religiosity: Issp 2008 data analysis. Aoyama J Cult Creat Stud 3:1–35

Manabe K (2012) Toward the integration of "theory" and "research" in the study of religion. Aoyama Stand J 7:283–298

Shye S (ed) (1978) Theory construction and data analysis in the behavioral sciences. Jossey-Bass social and behavioral science series. Jossey-Bass, San Francisco

Socio-economic and Gender Differences in Voluntary Participation in Japan

Miki Nakai

Abstract The aim of the present paper is to examine the relationship among participation in voluntary association, socio-economic position, and gender. Based upon a nationally representative sample in Japan in 2005 (N = 2,827), we classify a variety of voluntary organizations in terms of membership through which social integration/cohesion and social exclusion operate among different groups – such as social class, education, age, and gender. Correspondence analysis of voluntary participation data revealed that membership is differenciated by gender. It was confirmed that the distinction between old and new organizations seems to be valid in the Japanese context. The implications of the results are also discussed in terms of gender inequality and segregation not only in the economic arena but also in the society as a whole.

1 Introduction

A lot of social scientists regard voluntary organizations as mediators between the mass and the elite. Participation in voluntary associations has been seen as an agent by which weak individuals become strong and thereby has been regarded as 'social capital'. This 'social capital' concept has been studied by a number of social scientists; involvement in associational activity leads to the prosperity of democratic institutions (Putnam 2000).

At the same time, many of the early studies emphasized that voluntary associations were sorting mechanisms, that is to say, the groups divided and segregated people by their interests and ultimately by socio-economic, educational,

M. Nakai (✉)
Department of Social Sciences, College of Social Sciences, Ritsumeikan University,
56-1 Toji-in Kitamachi, Kyoto 603-8577 Japan
e-mail: mnakai@ss.ritsumei.ac.jp

W. Gaul et al. (eds.), *German-Japanese Interchange of Data Analysis Results*,
Studies in Classification, Data Analysis, and Knowledge Organization,
DOI 10.1007/978-3-319-01264-3_20,
© Springer International Publishing Switzerland 2014

and religious differences. Decades of research on voluntary association affiliation show how membership is related to various socio-demographic characteristics, such as social class, education, and age.

However, how various types of associational activities are classified has yet to be elucidated fully (Li et al. 2003; Nakai 2005). In this paper, we investigate the way how people's voluntary participation is segregated with a focus on gender as well as social class, education, and age. Some literature suggested that in many societies, gender is related to political activities; men are more active than women (Verba et al. 1978). It was also found that women were less engaged in unconventional forms of participation such as strikes and demonstrations. It has been suggested that the well-established gender gap in many common forms of political participation remained evident during the 1980s and early 1990s in many countries around the world (Barnes and Kaase 1979).

Some researchers, however, expect to find evidence that some of these gender differences have gradually diminished or even disappeared over time, with women becoming more active, especially among the younger generations in affluent modern societies (Inglehart and Norris 2003). For example, as more women enter the paid labour force, they may be less likely to join female-dominated groups which help to stabilize their personality and more likely to have voluntary participation patterns that are similar to men's patterns which enable them to earn their living and are useful for furthering one's career. Such arguments would fit nicely with evidence that there has been a reduction in gender segregation in some other domains in developed countries.

On the other hand, the validity of this 'reduction in gender segregation' perspective in the context of Japanese society is highly questioned, because the relative stability of occupational gender segregation and gender division of labour in households would lead to continued segregation of voluntary participation patterns. The persistently high level of gender segregation in the labour market, or paid work in Japan, has been investigated by several researchers (see e.g. Nakai 2009), but we know much less about the same phenomenon in voluntary associations, so-called unpaid social participation.

Therefore, the aim of this paper is to make visible social distinction between different types of voluntary organizations, so that we could assess the mechanisms through which social integration/cohesion and social exclusion operate in different groups, such as social class, education, age, and gender. Here, we report our preliminary classification of a variety of voluntary organizations in terms of membership, so that we can continue with further research to examine the role of the voluntary participation to facilitate trust, democracy, as well as gender equality in Japan.

1.1 Some Classification Systems for Voluntary Organizations

Some researchers have offered a classification system which groups various voluntary organizations (e.g. Li et al. 2003; Schofer and Fourcade-Gourinchas 2001).

For example, Li et al. (2003) employed a "working-class dominated" or "Labour" type of organization and a "service-class-dominated" or "Civic" type of organization to show that class difference in associational membership is a means for particular social groups to retain their exclusiveness by forming tight bonds with others similar to themselves. Homophily is often used in social network analysis to explain this effect (McPherson et al. 2001).

Meanwhile, Schofer and Fourcade-Gourinchas (2001) employed "new" versus "old" social movement association membership. They examined variation in civic involvement in associational activities among nations, as well as variation in the types of associational activities in which their citizens engaged.

These classification systems might be useful for their own arguments, but the types suggested are conceptualized as ideal types. They are a priori defined classifications, and for that reason, arbitrary. Also, the nature of the voluntary association in which people are involved may depend on the society they live in. Circumstances may be specific to each country under consideration. Country-specific institutional background (or the differences in popularity) may matter for the pattern of voluntary engagement. In this paper we will investigate how well the classification systems introduced in the literature work in a Japanese context. Based on the 2005 national survey of social stratification and mobility, we aim to classify various types of voluntary associations in terms of people's actual participation patterns or their membership.

2 Methods

To clarify the structural pattern of voluntary participation, correspondence analysis was adopted so that we get a multidimensional graphical visualization of the relationship among the categorical variables (Greenacre and Blasius 1994; Le Roux and Rouanet 2009).

2.1 Data

The data is from a nationally representative survey of social stratification and mobility conducted in 2005 in Japan. The sample has 2,827 respondents (1,343 men and 1,484 women), aged 20–69. The response rate of 44.2 % is considered as not very high. Younger males seem to be slightly under-represented. Respondents provided such socio-demographic information as age, education, occupation, as well as the involvement in a variety of associational activities. We analyze relations between social class and participation in the Japanese context.

2.2 Measurement of Respondents' Participation in Voluntary Organizations

Respondents' participation in a variety of civil society and community organizations are measured by the question of respondents' affiliation in each of 12 types of organizations/association groups as follows:

(1) Neighbour's associations **(NEIGHBR)**
(2) Community associations such as women's clubs, youth associations, senior's clubs, voluntary fire fighters' clubs **(COMMNTY)**
(3) Parent-teacher associations **(PTA)**
(4) Guilds such as agricultural cooperatives and fishery organizations, trade associations **(GUILD)**
(5) Trade unions **(UNION)**
(6) Political parties **(PARTY)**
(7) Supporters groups or fan clubs **(SUPPRT)**
(8) Local consumer cooperatives **(COOP)**
(9) Civic advocacy groups and NGOs **(NGO)**
(10) Voluntary groups **(VOLUNTR)**
(11) Religious associations **(RELIGIOUS)**
(12) Hobby and leisure clubs **(LEISURE)**

We can link these associations to the a priori classification systems by Li et al. (2003) and Schofer and Fourcade-Gourinchas (2001). Based on the framework of Li et al. (2003), "labour" type of engagement includes trade unions; "civic" type of engagement includes parent-teacher associations, political party, civic advocacy groups and NGOs, voluntary groups, religious associations, and hobby and leisure clubs. Based on the framework of Schofer and Fourcade-Gourinchas (2001), "new" social movement associations include local consumer cooperatives, civic advocacy groups and NGOs, and voluntary groups; "old" social movement associations include guilds, trade unions, and political party.

The words in bold in parentheses along with circle markers are used to represent each voluntary organization in Fig. 1.

2.3 Social Status Groups

It is of interest here to look at differences in voluntary organization involvement among social status groups and gender of respondents. Therefore respondents are grouped by occupation, education, age, and gender:

Respondent's occupational categories: (A) professional, (B) managerial, (C) clerical, (D) sales, (E) skilled manual, (F) semi-skilled manual, (G) non-skilled manual, (H) farm.

Respondent's employment status: (I) full-time, (J) part-time, (K) self-employed, (L) unemployed.
Respondent's education: (M) primary, (N) secondary, (O) tertiary (junior college), (P) tertiary (university).
Respondent's age: (Q) 20–29, (R) 30–39, (S) 40–49, (T) 50–59, (U) 60–69.

3   Results

3.1   Who Belongs to Voluntary Associations?

We first show socio-economic and gender differences in participation in Table 1. The rows correspond to the respondents' social status characteristics and the columns are the voluntary associations. Entries are the percentages of individuals claiming membership in each particular association. According to the overall percentages of the respondents who affiliate with each particular association, more than half of the respondents (54.9 %) are members of neighbour's associations. Only a fourth (24 %) did not belong to any voluntary association (not shown in table). Studies of comparative civic societies often tell that the United States is always at or near the top of volunteer participation, whereas Japan trails in the rear, and European countries such as Britain and France occupy the space in the middle. However, quite a number of people (3 out of 4) participate in at least one association. A recent comparative study of East Asia suggests that participation by Japanese is not at a low level, but relatively greater numbers of people are involved in various types of association than other east Asian countries and regions (Yoshino and Tsunoda 2010). The result here can be understood in the light of this argument.

There is some evidence that women are more likely than men to be involved in more associations. Women have an average of 1.67 memberships, whereas men average is 1.54 memberships (not shown in table). The difference in the average number of memberships is not big, but statistically significant ($p < 0.05$, using Student's t-test). One of the earliest studies of political behaviour in Western Europe and North America concluded that men exceeded women in the number of voluntary association memberships (e.g. Booth 1972). However, women are more active in contemporary Japan.

People who are in their 20s and unmarried participate in relatively less associations and, therefore, their social capital seems to be very limited.

3.2   Classification of Voluntary Association

We analyzed Table 1 by correspondence analysis. The first four singular values (with percentages of chi-square) are 0.30398 (44.36 %), 0.20362 (19.90 %), 0.16374

Table 1 Percentages of respondents who participate in 12 types of associations

	Percentages of membership in each association											
	(1)	(2)	(3)	(4)	(5)	(6)	(7)	(8)	(9)	(10)	(11)	(12)
All	54.9	17.1	13.1	11.0	8.9	1.6	4.1	9.1	1.9	6.6	4.8	26.0
Male subtotal	54.8	12.1	9.2	15.5	11.1	2.2	4.9	4.0	1.6	5.2	4.1	26.8
(A) Professional	55.9	8.4	22.4	7.7	17.5	0.0	2.8	8.4	3.5	4.9	11.2	34.3
(B) Managerial	70.5	11.6	16.1	18.8	12.5	8.0	9.8	7.1	2.7	8.9	2.7	35.7
(C) Clerical	55.0	11.1	14.0	13.5	20.5	2.9	5.3	3.5	1.2	3.5	2.9	27.5
(D) Sales	56.4	12.8	3.0	24.1	8.3	3.0	5.3	5.3	1.5	5.3	3.0	27.8
(E) Skilled manual	57.9	12.7	10.3	19.0	13.5	2.8	5.6	4.0	1.6	2.8	2.0	24.2
(F) Semi-skilled manual	49.7	18.5	9.3	10.6	16.6	0.0	2.6	3.3	1.3	2.6	6.6	23.2
(G) Non-skilled manual	36.4	10.6	1.5	9.1	6.1	1.5	1.5	1.5	1.5	4.5	3.0	22.7
(H) Farm	73.5	19.1	4.4	57.4	0.0	0.0	13.2	0.0	1.5	8.8	2.9	13.2
(I) Full-time	56.1	11.9	12.8	10.8	18.7	2.9	5.2	5.2	1.4	4.2	3.8	28.2
(J) Part-time	40.5	9.1	5.8	9.9	3.3	1.7	0.8	1.7	2.5	3.3	5.0	24.8
(K) Self-employed	69.2	18.5	8.1	47.9	0.5	0.9	8.5	3.3	2.8	6.6	5.7	22.3
(L) Unemployed	45.5	8.5	0.4	4.9	0.4	1.6	2.8	2.0	0.8	8.1	3.3	27.2
(M) Primary	55.1	10.1	2.2	18.9	2.6	4.0	4.8	1.3	0.9	1.8	3.5	12.8
(N) Secondary	56.0	13.6	8.3	17.4	12.3	1.5	5.2	3.6	1.0	4.2	3.8	27.1
(O) Tertiary (junior college)	65.7	14.3	14.3	14.3	20.0	5.7	8.6	5.7	0.0	17.1	5.7	37.1
(P) Tertiary (university)	51.4	10.2	14.2	10.2	13.0	2.3	4.1	6.1	3.3	7.9	4.8	33.3
(Q) 20–29	10.9	5.8	0.6	5.1	8.3	0.0	0.6	0.6	1.3	2.6	1.9	28.8
(R) 30–39	37.2	10.5	13.0	6.9	19.0	1.2	2.8	3.2	0.8	2.4	2.0	25.5
(S) 40–49	63.3	20.0	23.3	13.7	18.9	0.7	5.2	7.4	1.9	4.4	8.1	27.8
(T) 50–59	70.9	9.8	7.8	21.6	10.8	3.3	6.2	5.9	1.0	5.6	3.6	24.8
(U) 60–69	65.5	11.8	0.8	21.9	1.4	4.1	6.8	1.9	2.7	8.5	3.8	27.7
Female subtotal	55.0	21.6	16.6	7.0	6.9	1.0	3.4	13.7	2.2	7.9	5.4	25.2
(A) Professional	56.1	24.2	31.8	3.2	22.3	1.3	1.9	20.4	3.8	10.2	8.3	27.4
(B) Managerial	61.5	38.5	7.7	46.2	15.4	0.0	15.4	30.8	0.0	7.7	0.0	27.4
(C) Clerical	45.1	17.8	20.8	8.3	12.5	1.1	3.0	14.4	2.7	6.4	6.4	28.0
(D) Sales	56.6	17.8	17.8	14.7	6.2	0.0	0.0	10.1	0.8	3.1	2.3	17.1
(E) Skilled manual	52.1	23.4	18.1	7.4	6.4	0.0	5.3	10.6	2.1	3.2	5.3	16.0
(F) Semi-skilled manual	62.2	21.6	14.4	6.3	6.3	0.9	0.9	5.4	0.0	4.5	5.4	14.4
(G) Non-skilled manual	54.5	20.0	15.5	5.5	4.5	1.8	3.6	15.5	1.8	12.7	10.0	24.5
(H) Farm	50.0	36.0	6.0	34.0	0.0	2.0	6.0	8.0	0.0	8.0	0.0	24.0
(I) Full-time	45.9	17.1	18.5	7.6	22.4	1.1	2.2	12.3	2.5	5.3	5.3	22.7
(J) Part-time	58.1	20.7	24.0	5.4	3.8	0.9	2.6	13.6	1.6	7.3	7.5	21.9
(K) Self-employed	55.8	34.0	10.9	26.5	0.7	0.7	4.8	15.6	1.4	8.8	2.0	27.2
(L) Unemployed	58.3	21.8	11.4	2.7	0.9	1.1	4.5	14.1	2.7	9.6	4.5	28.9
(M) Primary	50.8	20.2	2.0	10.1	2.4	1.6	5.6	4.0	1.6	6.5	4.0	18.1
(N) Secondary	58.2	22.0	16.5	7.2	5.9	1.0	3.4	13.3	2.5	7.6	6.3	27.3
(O) Tertiary (junior college)	53.5	25.2	30.2	5.4	12.9	0.5	3.0	21.8	1.5	8.9	3.0	25.7
(P) Tertiary (university)	45.2	17.4	23.2	3.2	11.6	0.6	0.6	20.6	2.6	10.3	5.8	23.9
(Q) 20–29	14.2	1.2	0.6	0.0	8.3	0.6	1.2	4.1	0.6	3.6	4.7	16.6
(R) 30–39	44.6	23.4	33.8	1.9	13.0	0.7	1.1	13.4	0.7	3.7	3.3	17.8
(S) 40–49	65.1	33.9	45.2	8.6	9.2	0.7	1.4	21.9	2.4	7.2	7.9	24.7
(T) 50–59	62.2	15.9	6.1	10.3	6.3	1.3	4.0	15.3	2.9	8.7	5.6	25.4
(U) 60–69	65.7	25.8	0.0	9.3	0.5	1.3	7.2	10.1	3.2	12.5	5.1	34.6

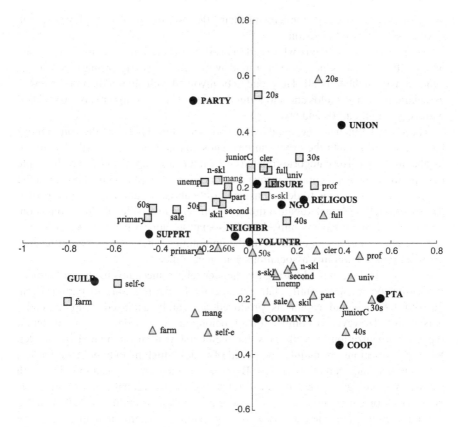

Fig. 1 Structure of participation, social status and gender: Dimension 1 × Dimension 2

(12.87 %), and 0.12902 (7.99 %). Therefore we can restrict attention to the first two dimensions following the elbow criterion. These two dimensions display 64.3 % of the total dependence between the row and the column variables.

Then, we used the configuration space shown in Fig. 1 to map voluntary organizations, respondents' socio-economic status, educational level, age and gender which allowed us to identify social and gender boundaries. The figure represents the positions of the voluntary associations by circle markers, social groups of males by square markers and social groups of females by triangle markers in two-dimensional space.

Inspecting Fig. 1, we see that the first dimension (horizontal axis) contrasts the urban new middle-class types of association in the positive part from the old middle-class types of association in the negative part. White-collar employees and civil servants, as well as respondents with a university degree are in the right half and they tend to participate in new middle-class type organizations such as trade unions, PTAs (parent-teacher associations), local COOPs (consumer cooperatives), and NGOs. Moreover, respondents in their 20s and 30s are also displayed on the positive side of this dimension.

By contrast, respondents who work as self-employed and persons doing home handicraft, as well as people who found work after finishing primary and lower education are in the left half. They tend to be involved with old middle-class types of associations such as guilds and supporters' groups. These groups are also associated with people in their 50s and 60s.

The second dimension (vertical axis) mirrors a contrast between the community-based associations and the organizations whose basis is not necessarily located in local community. The categories related to the interest groups through which people have political involvement or improve the condition of employment show up in the upper part. They include trade unions and political parties and have been regarded as "traditional" political organizations. On the negative side of this dimension, on the other hand, the community-based associations appear that are located in and provide services to neighbourhoods and communities. They include parent-teacher organizations, community associations, and many others.

Much of the literature almost ignore gender related mechanisms that are seen as producing social capital until relatively recently. Social relations, or participation, through which social capital is produced are rarely analysed from a gender perspective. However, a comparison of belonging to a wide range of different types of organizations reveals how far membership is differentiated by gender. Some organizations are mainly composed of males which include political parties, trade unions, and recreation circles. By contrast, there are associations in which women predominate, especially those related to the traditional role of women as housewives or caregivers, such as those concerned with child and safeguarding consumer rights and interests, as well as providing women's human rights. The distinction between "old" and "new" organizations combined with gender proved to be valid in this empirical analysis. However, this new/old distinction can be conceptualized in different way from some previous researchers such as Schofer and Fourcade-Gourinchas (2001), where "new" social movements focus on issues related to human rights and cover associations which have developed rapidly since the late 1960s, whereas old social movement associations aim to gain access for the working class to economic wellbeing. Instead, our result shows that the distinction between new middle-class types and old middle-class types of associations can be useful in the differentiation of associations in a Japanese context. Conventional forms of political participation such as parties and trade unions are shown to be associated with males, whereas women are active in the alternative channels such as NGOs and grass-roots voluntary associations rather than via traditional modes of political expression. Having said that, participation patterns of young women working on a full-time basis appear to be similar to that of male white-collar employees from the viewpoint of types of associations.

In this figure, most of the points of male involvement with square markers are on the positive part of dimension 2, with two exceptions: male respondents who are self-employed or farmers. On the other side of dimension 2, almost all the points of female involvement marked with triangle markers show up, with

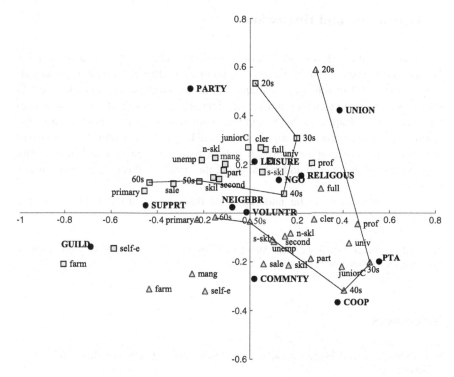

Fig. 2 Horseshoe effect. The plot of the age categories shows an arching (horseshoe shaping) pattern

two exceptions: women in their 20s or women those who work full-time. Therefore, women in their 20s and women who work full-time are on the same side as men in terms of dimension 2, which suggests that they show the same degree of participation in paid work as their male counterparts and, at the same time, tend to show associational membership patterns similar to men's patterns.

The age group variable forms a horseshoe, a typical structure we usually find in ordered categorical data (see Fig. 2). We also found that each pair of the same age groups of men and women has a parallel curve, but their participation pattern is completely different through their 30s and 40s. This result implies that segregation exists which reflects existing patterns of gender inequality not only in the labour force but also in society as a whole. Gender division of labour becomes well established when men and women are in their 30s. This segregation might harm women's full participation in society.

4 Conclusion and Discussion

The main findings of this study are as follows: This study confirms that the distinction between old middle-class type and new middle-class type organizations seems to be valid in this empirical analysis in the Japanese context. We see that women's voluntary participation may be changing as a result of changing gender roles. However, despite the movement of women's work from household to the labour market, participation is highly segregated by gender in Japan. These analyses show that the social space approach with a focus on gender can bring fruitful information for the overall understanding of the current status of gender inequality in Japan.

Further research must be undertaken in these areas utilizing the data of a follow-up study.

Acknowledgements I thank the Social Stratification and Social Mobility (SSM) 2005 Committee providing the data set used in this study.

References

Barnes SH, Kaase M (1979) Political action: mass participation in five Western democracies. Sage, Beverly Hills

Booth A (1972) Sex and social participation. Am Sociol Rev 37(2):183–193

Greenacre MJ, Blasius J (1994) Correspondence analysis in the social sciences: recent developments and applications. Academic, London/San Diego

Inglehart R, Norris P (2003) Rising tide: gender equality and cultural change around the world. Cambridge University Press, Cambridge

Le Roux BL, Rouanet H (2009) Multiple correspondence analysis. Sage, Thousand Oaks

Li Y, Savage M, Pickles A (2003) Social capital and social exclusion in England and Wales (1972–1999). Br J Sociol 54(4):497–526

McPherson M, Smith-Lovin L, Cook JM (2001) Birds of a feather: homophily in social networks. Ann Rev Sociol 27(1):415–444

Nakai M (2005) Social stratification and social participation: the role of voluntary association membership. In: Ojima F (ed) Research on gender and social stratification in contemporary Japan. Report of Grant-in-Aid for Scientific Research (B)(1), Ministry of Education, Culture, Sports, Science and Technology, pp 53–63

Nakai M (2009) Occupational segregation and opportunities for career advancement over the life course. Jpn Sociol Rev 59(4):699–715

Putnam RD (2000) Bowling alone: the collapse and revival of American community. Simon & Schuster, New York

Schofer E, Fourcade-Gourinchas M (2001) The structural contexts of civic engagement: voluntary association membership in comparative perspective. Am Sociol Rev 66(6):806–828

Verba S, Nie NH, Kim JO (1978) Participation and political equality: a seven-nation comparison. Cambridge University Press, Cambridge

Yoshino R, Tsunoda H (2010) A note on social capital: from a viewpoint of cross-national comparative methodology. Jpn J Behaviormetr 37(1):3–17

Estimating True Ratings from Online Consumer Reviews

Diana Schindler, Lars Lüpke, and Reinhold Decker

Abstract Online consumer reviews have emerged in the last decade as a promising starting point for monitoring and analyzing individual opinions about products and services. Especially the corresponding "star" ratings are frequently used by marketing researchers to address various aspects of electronic word-of-mouth (eWOM). But there also exist several studies which raise doubts about the general reliability of posted ratings. Against this background, we introduce a new framework based on the Beta Binomial True Intentions Model suggested by Morrison (J Mark Res 43(2):65–74, 1979) to accommodate the possible uncertainty inherent in the ratings contained in online consumer reviews. We show that, under certain conditions, the suggested framework is suitable to estimate "true" ratings from posted ones which proves advantageous in the case of rating-based predictions, e.g. with respect to the willingness to recommend a product or service. The theoretical considerations are illustrated by means of synthetic and real data.

1 Introduction

Today, the Internet has established itself as an important channel for online trading, purchasing and chatting. Consumers intensively use new forms of communication such as online forums, blogs, discussion groups as well as online review platforms like those offered by *Amazon.com*, *Epinions.com* or *Airlinequality.com* to express their opinions and thoughts about products or services. Hence, the respective consumer feedbacks have emerged as an important database for marketing research.

D. Schindler · L. Lüpke · R. Decker (✉)
Department of Business Administration and Economics, Bielefeld University,
Bielefeld, Germany
e-mail: dschindler@uni-bielefeld.de; lluepke@uni-bielefeld.de; rdecker@uni-bielefeld.de

W. Gaul et al. (eds.), *German-Japanese Interchange of Data Analysis Results*,
Studies in Classification, Data Analysis, and Knowledge Organization,
DOI 10.1007/978-3-319-01264-3_21,
© Springer International Publishing Switzerland 2014

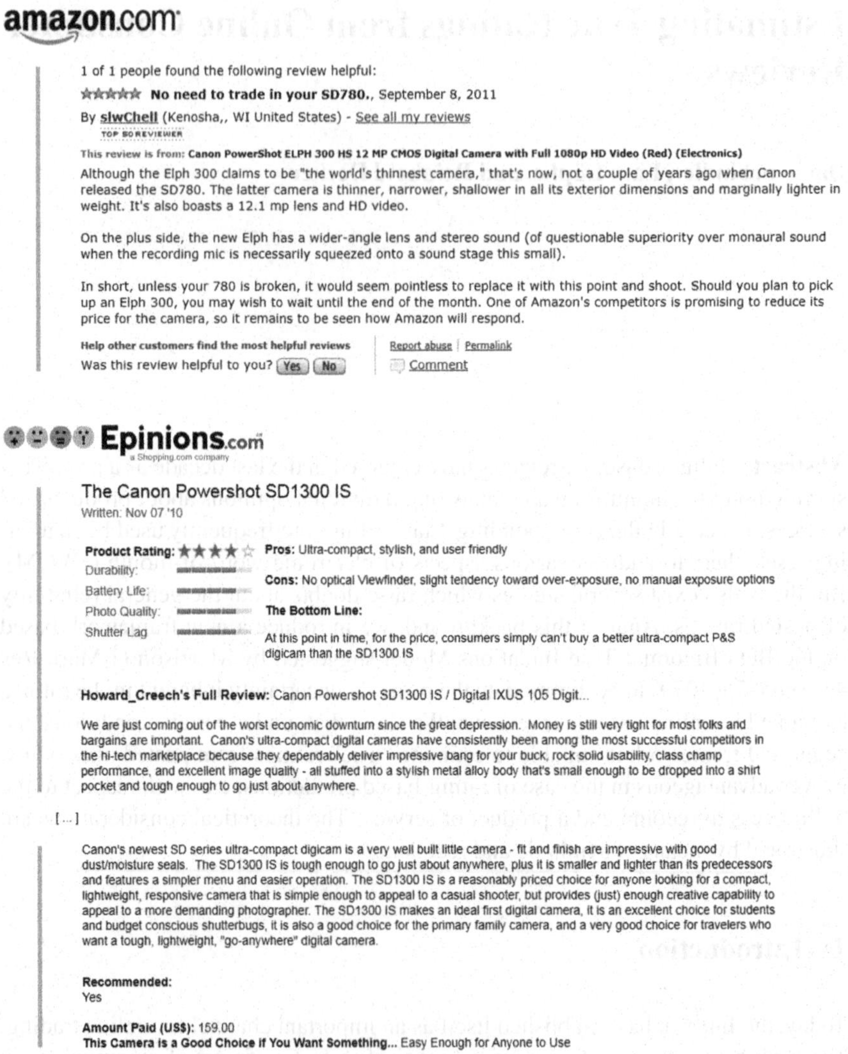

amazon.com

⊕☺☻☺⊕ **Epinions**.com
a Shopping.com company

Fig. 1 Examples of online consumer reviews from *Amazon.com* and *Epinions.com*

The structure of the reviews can vary significantly depending on the platform on which the reviews are posted. Two very popular formats are shown in Fig. 1.

Online consumer reviews typically contain so-called "star" ratings (e.g., expressed on a 5-point scale) and different formats of full texts. In some cases, online consumer reviews also include additional information about pros and cons and the recommendation or non-recommendation of the respective product or service as, see, e.g., online consumer reviews from *Epinions.com*.

Despite this broad range of information sources, studies in this context are often based on data sets which only contain "star" ratings and texts as illustrated by the *Amazon.com* example. An obvious characteristic of those reviews is the fact that the two components differ greatly regarding their data analytic complexity. The difficulties with respect to the statistical analysis of texts may be one reason why many studies are based on rating information instead of texts. The corresponding aims of scientific investigations vary considerably. There are studies which analyze the value of online consumer reviews for firms by evaluating posted ratings (Dellarocas et al. 2004). Other papers analyze the value of online ratings with respect to the development of marketing strategies (Chen et al. 2003; Thomas et al. 2011; Lee et al. 2011) or assess the value of online consumer reviews for the economy in general (Jiang and Chen 2007). Based on posted ratings, Dellarocas et al. (2010) recently investigated whether consumer reviews are primarily written for popular "hit" products or for "niche" products. Furthermore, there is a growing number of studies devoted to the quantification or measurement of the impact of posted ratings on (future) sales (Chen et al. 2004; Godes and Mayzlin 2004; Chevalier and Mayzlin 2006; Joeckel 2007; Moe and Trusov 2011).

While many researchers are directing their efforts to the analysis of online consumer ratings for marketing purposes, there is also a growing research devoted to the question whether these ratings are a consistent marketing database at all. Actually, several aspects are conceivable to be suspicious regarding the significance of posted consumer ratings when they are used as the only source of information. For example, Li and Hitt (2008), as well as Moe and Schweidel (2011), confirm that reviewers' purchase intentions may vary over time because of other reviews. Hence, later reviews may be influenced by earlier ones. Moe and Trusov (2011) further support this empirically, which indicates that posted ratings are not necessarily robust over time and cannot inevitably be interpreted as the true quantifications of opinions. Li and Hitt (2010), furthermore, argue that posted ratings on the aggregate level do not necessarily reflect the objective value or quality of a product or service. Most recently, Hao et al. (2011) provided further evidence in this respect by studying consumer ratings in third-party software application markets. Hu et al. (2006) falsified the assumption that average ratings calculated from online consumer reviews fully reflect aggregate preferences. Instead, they pointed out that average ratings only represent a compromise of extreme opinions. Later, Hu et al. (2009) state that online consumer reviews are often written in a rather positive than objective way (they categorized this as J-shaped distribution) and, as a consequence thereof, are biased and cannot be taken per se as a consistent data base. In addition, they point out that online reviewers either tend to report their distinct satisfaction or distinct dissatisfaction regarding a purchased product. This means that people who are neither extremely satisfied nor extremely unsatisfied typically have little motivation to write about this "average feeling". From the aforementioned study it can be concluded that online consumer reviews coming along with a 1-star (minimum) or a 5-star (maximum) rating may be overrepresented, whereas those with ratings in between may be underrepresented. Both together would remarkably influence the results of data analyses based thereon. The results of Hu et al. (2009)

further substantiate the assumption that analyzing posted "star" ratings without examining the reliability of the corresponding frequency distribution may lead to erroneous interpretations. Finally, Tsang and Prendergast (2009) empirically evaluated the relationship between review texts and numerical ratings and found that the trust in online consumer reviews strongly depends on the consistency of textual evaluation and posted rating (e.g. positive ratings and positive valences of free texts). Especially, in consumer-to-consumer relationships it is important that text valences and ratings correlate, which can be seen as an implicit indicator of consistency.

The argument of Abulaish et al. (2009) points to a similar direction and suggest examining both review texts and rating information and, hence, managing possibly existing inconsistencies in the posted ratings that way. They argue that, basically, the whole review text has to be read to be able to correctly interpret the true meaning of the given rating. Even though there are several activities, e.g., with respect to the use of text mining tools to aggregate the textual parts of online consumer reviews, such that one does not have to read all the texts available for a product or service of interest, reviewers' subjectivity still remains a problem for review data analysis (Abulaish et al. 2009; Liu et al. 2005; Liu 2010).

The consolidation of the above-reported streams of research directly leads to the following implication: When analyzing online consumer reviews on the aggregate level and only using rating information then it cannot be excluded that the results are biased, either because of a J-shaped distribution of the ratings (Hu et al. 2009) or because of reviewers' subjectivity and situational influences (e.g. the basic mood of the reviewer when writing the review). Posted, i.e. observed, ratings can therefore be interpreted as the outcome of a stochastic process. Against this background, the aim of this paper is to introduce and illustrate an easy-to-use framework, which helps to analyze and interpret inconsistencies that result from the above-mentioned conditions.

The remainder of the paper is organized as follows. In Sect. 2, we provide a brief sketch of the Beta Binomial True Intentions Model by Morrison (1979) (Sect. 2.1), introduce an adapted version to deal with ratings (Sect. 2.2), and finally apply it to a synthetic data set (Sect. 2.3), which represents different rating distributions. In Sect. 3, the new approach is applied to two real data sets generated from online consumer reviews in order to demonstrate its basic functionality and to shed light on some important aspects of its practical use. Section 4 discusses the results and corresponding limitations and outlines directions for future research.

2 Theoretical Framework

2.1 Beta Binomial True Intentions Model (BBTIM)

More than three decades ago, Morrison (1979) suggested and empirically verified a simple stochastic framework for transforming stated purchase intentions into estimated "true" – but unknown – purchase intentions, which can then be used to estimate future purchase probabilities. The starting point of this so-called True Intentions Model (TIM) is the assumption that, whenever a person is asked whether s/he intends to buy a product in the future, the stated intention may change over time: Let us assume, for example, that a person, who is asked today whether s/he will buy a dishwasher during the next year, would answer this question with "no". In this case, the stated purchase intention clearly points to a purchase probability close or even equal to zero. But then, 6 months later, the respondent's present dishwasher breaks down. Now two scenarios are imaginable: (1) the respondent will repair it or (2) s/he will buy a new one. The latter, of course, would contradict the previously stated intention.

To deal with this kind of information instability Morrison (1979) suggested the TIM and used it to transform stated (i.e. empirically observed) intentions (I_s) into estimates of the unknown "true" intentions (I_t). The key idea behind this very elegant approach was to "deflate high stated intention and inflate low stated intention" (Morrison 1979, p. 66) in order to get a more realistic picture of the "true" intentions existing in the product category of interest. Therefore, he first defined the following simple linear relationship

$$I_t = a + bI_s \tag{1}$$

with unknown parameters a and b ($a, b > 0$ and $a + b < 1$) to be estimated from survey data. For the unknown parameters a and b the underlying "regression to the mean" principle requires that $a, b > 0$ and $a + b < 1$ (Healy and Goldstein 1978). Both intention variables are supposed to be linearly dependent on each other. Furthermore, it is assumed that a respondent answers on an $(n + 1)$-point scale with levels $0, 1, 2, \ldots, n$ (where 0 means definitely no purchase intention and n points to a definite purchase intention). Expressed in a more formal way, we get

$$P(I_s = r; n, I_t) = \binom{n}{r} I_t^r (1 - I_t)^{n-r}, \text{ with } r = 0, 1, \ldots, n \text{ and } I_t \in (0, 1) \tag{2}$$

where I_s can be interpreted as the outcome of scale level based Bernoulli draws and therefore can be assumed to follow a binomial distribution with parameters n and $I_t \in (0, 1)$.

Following this, the estimation of I_t is based on

$$g(I_t; \beta_1, \beta_2) = \frac{\Gamma(\beta_1 + \beta_2)}{\Gamma(\beta_1)\Gamma(\beta_2)} I_t^{\beta_1 - 1} (1 - I_t)^{\beta_2 - 1}. \tag{3}$$

Starting from the above probability functions and by assuming the "true" purchase intentions I_t to be beta-distributed, Morrison (1979, p. 67) developed his Beta Binomial True Intentions Model (BBTIM), which provides the methodological guide for our own approach:

$$P(I_s = r; n, \beta_1, \beta_2) = \binom{n}{r} \frac{B(\beta_1 + r, \beta_2 + n - r)}{B(\beta_1, \beta_2)}, \tag{4}$$

$$\text{with } r = 0, 1, \ldots, n \text{ and } B(\beta_1, \beta_2) = \frac{\Gamma(\beta_1)\Gamma(\beta_2)}{\Gamma(\beta_1 + \beta_2)}.$$

BBTIM provides a methodologically elegant way to deal with information instability that might occur due to circumstances respondents do not or cannot anticipate when asked about their purchase intentions.

In the case of online consumer reviews we have got a similar situation: Reviewers often post their opinions about a product or service under circumstances that are unknown to the marketing researcher. In addition, various situational factors might influence the rating process (see, e.g., Hao et al. 2011) or the underlying opinion changes over time (see, e.g., Moe and Schweidel 2011). Actually, the posted opinions can be the outcome of an actual purchase, as well as the result of past experiences, and they can be influenced by the current mood and/or the missing willingness to carefully balance the exact rating the product or service deserves. Inconsistency may also result from the non-willingness to report "average feelings" (see above) or the willingness to only write about extremely good or bad experiences. From a marketing researcher's perspective, all these factors suggest the interpretation of the posted (observed) ratings as the outcome of a latent stochastic process.

2.2 Beta Binominal True Ratings Model (BBTRM)

Analogous to Morrison's use of the stated intention I_s, in the case of online consumer reviews, Y^{posted} denotes the posted "star" rating with realizations $r = 0, 1, 2, \ldots, R$, whereas Y^{true} refers to the unknown "true" rating, which has to be estimated from posted/observed ratings. According to Morrison (1979) the following model development starts from two basic assumptions:

A_1:   The unknown "true" ratings are a linear function of the posted ones (i.e., posted rating = "true" rating + error).
A_2:   The unknown "true" ratings vary across the reviews posted for a particular product or service.

Furthermore, it is hypothesized:

H_1:   The variation of the "true" ratings can be described by a beta distribution.
H_2:   The "true" ratings can be estimated from the posted ones.

Assumption A_1 leads to the simple relationship

$$Y^{true} = a + bY^{posted}. \tag{5}$$

The posted ratings are interpreted in the following as individual responses on an $(R + 1)$-point rating scale with levels $r = 0, 1, 2, \ldots, R$. A value equal to zero indicates that the reviewer has a clearly negative opinion about the respective product or service, whereas the highest level R points to a clearly positive opinion about the rated object. The observed realizations r of the random variable Y^{posted} can then be interpreted as the outcome of a binomial process with parameters R and unknown probability Y^{true}:

$$P(Y^{posted} = r; R, Y^{true}) = \binom{R}{r} (Y^{true})^r (1 - Y^{true})^{R-r}, \tag{6}$$

$$\text{with } r = 0, 1, \ldots, R \text{ and } Y^{true} \in (0, 1).$$

Continuing with assumption A_2 we consider the "true" rating Y^{true} to follow a beta distribution with parameters $\beta_1, \beta_2 > 0$. The probability of observing a particular rating r then takes the form:

$$P(Y^{posted} = r; R, \beta_1, \beta_2) = \binom{R}{r} \frac{\Gamma(\beta_1 + \beta_2)}{\Gamma(\beta_1)\Gamma(\beta_2)} \frac{\Gamma(r + \beta_1)\Gamma(R - r + \beta_2)}{\Gamma(R + \beta_1 + \beta_2)}. \tag{7}$$

The unknown parameters β_1 and β_2 can be estimated from the observed rating data by means of the maximum likelihood method. According to Kalwani and Silk (1982, p. 284) the regression parameters a and b of Eq. (5) can be deduced from the conditional expectation as follows:

$$E(Y^{true} = y^{true} | Y^{posted} = r) = \frac{r + \beta_1}{R + \beta_1 + \beta_2} = \frac{\beta_1}{R + \beta_1 + \beta_2} + \frac{1}{R + \beta_1 + \beta_2} r. \tag{8}$$

The empirical counterpart of Eq. (5), providing an estimator $\hat{y}^{true}$ for the "true" rating, then reads

$$\hat{y}^{true} = \hat{a} + \hat{b}r, \text{ with } \hat{a} = \frac{\hat{\beta}_1}{R + \hat{\beta}_1 + \hat{\beta}_2} \text{ and } \hat{b} = \frac{1}{R + \hat{\beta}_1 + \hat{\beta}_2}. \tag{9}$$

The above framework is referred to as the Beta Binomial True Ratings Model (BBTRM) in the following.

2.3 Application of the BBTRM Framework to Synthetic Data

In order to verify hypotheses H_1 and H_2 several data sets with varyingly distributed ratings are considered. As already mentioned the "star" ratings contained in online consumer reviews often have a J-shaped distribution (see, e.g., Chevalier and

Table 1 Illustration of different frequency distributions

$r = Y^{posted}$	$n_{Y^{posted}}$ [absolute]			$n_{Y^{posted}}$ [relative]		
	Example 1	Example 2	Example 3	Example 1	Example 2	Example 3
0	102	120	70	0.20	0.24	0.14
1	96	95	30	0.19	0.19	0.06
2	96	70	40	0.19	0.14	0.08
3	102	95	125	0.20	0.19	0.25
4 (=R)	104	120	235	0.21	0.24	0.47
n^{posted}	500	500	500	1	1	1

Table 2 Parameter estimates for the synthetic data sets

	Example 1	Example 2	Example 3
$\hat{\beta}_1$	0.95	0.66	0.77
$\hat{\beta}_2$	0.93	0.66	0.33
$\hat{a}$	0.16	0.12	0.15
$\hat{b}$	0.17	0.19	0.20
χ^{2emp}	0.16	2.94	39.47

(rounded values)

Mayzlin 2006; Hu et al. 2009; Li and Hitt 2008), indicating a strong positive opinion about the respective product or service. Own comprehensive (work-in-progress) analyses of different product categories underscore this appraisal. Besides this, uniform and U-shaped frequency distributions of overall ratings can be observed in numerous categories. We therefore created three synthetic data sets which illustrate these common distributions and demonstrate the application of BBTRM in general with these data sets.

Example 1 approximates a *uniform distribution*, Example 2 a *U-shaped distribution* and Example 3 represents a *J-shaped distribution*. All three data sets are based on $n^{posted} = 500$ simulated rating frequencies. Table 1 displays the test data sets.

In the case of Example 1 the maximum likelihood estimation leads to $\hat{\beta}_1^{Example\ 1} = 0.95$ and $\hat{\beta}_2^{Example\ 1} = 0.93$. As a result, we get $\hat{y}^{true,Example\ 1} = \frac{r+0.95}{4+0.95+0.93}$ and $P^{Example\ 1}(r = 0; 4, 0.95, 0.93) = \binom{4}{0} \frac{\Gamma(0.95+0.93)}{\Gamma(0.95)\Gamma(0.93)} \frac{\Gamma(0+0.95)\Gamma(4-0+0.93)}{\Gamma(4+0.95+0.93)} = 0.202$ (see Eq. (7)). The probabilities $P(r)$ for $r = 1, 2, 3, 4$ are calculated the same way. Table 3 shows the probabilities $P(r)$ and the resulting estimated frequencies $\hat{n}_{Y^{true}}$ for all three examples. The corresponding estimates $\hat{\beta}_1$ and $\hat{\beta}_2$, as well as $\hat{a}$ and $\hat{b}$ (see Eq. (9)), are shown in Table 2.

The results in Tables 2 and 3 show that the available BBTRM parameter estimates provide acceptable approximations $\hat{n}_{Y^{true}}$ of the observed frequencies $n_{Y^{posted}}$. To further verify this impression χ^2-tests were carried out (H_0: "*The observed frequencies $n_{Y^{posted}}$ of the posted ratings equal their estimated counterparts $\hat{n}_{Y^{true}}$*."). The non-rejection of this hypothesis implies that hypotheses H_1 and H_2 (see Sect. 2.2) cannot be rejected as well. In the present case H_1 and H_2 cannot be rejected for Examples 1 and 2, but have to be rejected for Example 3 at the 1 % level.

Table 3 Results for the synthetic data sets

r	$P(r)$			$\hat{n}_{Y^{true}} = P(r)\cdot n^{posted}$		
	Example 1	Example 2	Example 3	Example 1	Example 2	Example 3
0	0.202	0.243	0.116	$101 = 0.202\cdot500$	121	58
1	0.196	0.175	0.107	$98 = 0.196\cdot500$	88	54
2	0.194	0.164	0.122	$97 = 0.194\cdot500$	82	61
3	0.198	0.175	0.170	$99 = 0.198\cdot500$	88	85
4	0.210	0.243	0.485	$105 = 0.210\cdot500$	121	242

(rounded values)

Accordingly, in the first two cases, the assumption that the "true" product ratings follow a beta distribution cannot be falsified. Consequently, the above examples allow the preliminary conclusion that the BBTRM framework should work in the case of approximately U-shaped or uniformly shaped frequency distributions. We will discuss this in more detail in Sect. 4.

3 Estimation of "True" Ratings Using the BBTRM Framework

3.1 Application of the BBTRM Framework to Real World Data Sets

In the following, the BBTRM framework is applied to two real world data sets from fields where online reviewing has been popular for a number of years. The first data set was crawled from *Airlinequality.com*, the "world's leading airline and airport review site", the second one represents customer opinions from *Apartmentratings.com*, the "leading source of apartment reviews by renters". *Airlinequality.com* reviews can be assumed to be based primarily on the experiences typically resulting from the last flight with the respective airline, normally offering the main service for a couple of hours. In contrast to this, *Apartmentratings.com* reviews can be assumed to typically reflect the impressions and experiences gained over months or even years while living in the respective apartment. These different characteristics have a significant influence on the "true" ratings as will be shown later.

The first data set contains in total 491 ratings for four randomly selected airlines. In order to build a bridge to the widely used 5-point rating scales and to achieve a better allocation of the rating frequencies the original 10-point scale was transformed into a 5-point scale with rating groupings $(0; 1; 2)\hat{=}r=0$, $(3; 4)\hat{=}r=1$, $(5)\hat{=}r=2$, $(6; 7)\hat{=}r=3$ and $(8; 9; 10)\hat{=}r=4$ (see Juster (1966) and Morrison (1979) for similar aggregations). Neutral labels "Airline 1", "Airline 2", etc. are used for data privacy reasons. Table 4 illustrates the data used.

Table 4 Airline-specific frequencies of posted ratings

r	$n_{\gamma posted}$ Airline 1	Airline 2	Airline 3	Airline 4
0	39	63	54	19
1	18	12	19	6
2	10	16	6	7
3	24	14	15	15
4	46	14	31	63
n^{posted}	137	119	125	110

Table 5 BBTRM results for the airline data

	Airline 1	Airline 2	Airline 3	Airline 4
$\hat{\beta}_1$	0.38	0.25	0.23	0.37
$\hat{\beta}_2$	0.33	0.61	0.34	0.15
$\hat{a}$	0.08	0.05	0.05	0.08
$\hat{b}$	0.21	0.21	0.22	0.22
	$\hat{n}_{\gamma true} = P(r) \cdot n^{posted}$			
r	Airline 1	Airline 2	Airline 3	Airline 4
0	38	62	54	18
1	17	17	15	9
2	16	12	12	8
3	19	12	13	11
4	47	15	31	64
$\hat{n}_{true}$	137	118	125	110
H_0	The observed frequencies $n_{\gamma posted}$ of the posted ratings equal their estimated counterparts $\hat{n}_{\gamma true}$. ($\alpha = 0.01$; $\chi^{2crit} = 9.21$; $df = 2$)			
χ^{2emp}	3.66	3.29	4.26	2.21
H_0	*Not rejected*	*Not rejected*	*Not rejected*	*Not rejected*
Conclusions	The variation of the "true" ratings can be described by a beta distribution (H_1). The "true" ratings can be estimated from the posted ones (H_2).			

(rounded values)

Information about recommendation frequencies was available as well. We refer to this in Table 8 ($n_{rec^{posted}}$). The total number of reviews and ratings for Airline 1 to Airline 4 vary between 137 and 110, respectively, which meets usual sample sizes in this field of application at the time of data collection.

The airlines considered can be distinguished from each other on the basis of their different frequency distributions. The distribution of Airline 1 is approximately U-shaped, whereas the ratings for Airline 2 are rather negative. The data structure of Airline 3, in a sense, is like a mixture of that of Airline 1 and Airline 2. Finally, the frequency distribution of Airline 4 tends to be approximately J-shaped. Insofar the selected data sets are heterogeneous enough to build a meaningful database for model testing. Table 5 shows the results we get when applying BBTRM to this data.

Table 6 City-specific
frequencies of posted
apartment ratings

		$n_{Y\,posted}$		
r		City 1	City 2	City 3
0		448	2,337	441
1		259	1,399	227
2		267	1,294	227
3		197	1,193	199
4		241	1,257	279
n^{posted}		1,412	7,480	1,373

In all cases the "true" ratings could be estimated at a satisfactory level and, accordingly, accounting for unobservable influences (that might have occurred during the rating process) was possible (see the results of χ^2-tests in Table 5). Estimated $\hat{\beta}$ parameters as well as the corresponding function parameters $\hat{a}$ and $\hat{b}$ are also given in Table 5.

Additionally, the robustness of the available estimates was tested. Therefore, each sample was reduced randomly 10 times and the resulting parameter estimates $\hat{\beta}_1$, $\hat{\beta}_2$, $\hat{a}$ and $\hat{b}$ were compared. On average, the values of $\hat{\beta}_1$ vary by approximately 1 % in the case of data set 1 (U-shaped data), 4 % in the case of data set 2 (predominantly negative), 2 % in the case of data set 3 (mixture of data sets 1 and 2) and 8 % in the case of data set 4 (J-shaped data). The corresponding variations of $\hat{\beta}_2$ are 2, 2, 0.3 and 0.7 %. In the case of data sets 1 and 2 $\hat{a}$ and $\hat{b}$ do not differ notably (<0.5 %). This effect also occurs for the $\hat{b}$ parameters of data sets 3 and 4. The $\hat{a}$ parameters vary by 2 % for data set 3 and 12 % for data set 4. All in all, the respective variations tend to be mostly minor, which indicates acceptable robustness of the predictions. The results of χ^2-testing further support this. Accordingly, H_0 ("*The observed frequencies $n_{Y\,posted}$ of the posted ratings equal their estimated counterparts.*") cannot be rejected in any of the cases considered.

In order to further substantiate these findings the model was also applied to the above-mentioned apartment ratings (see Table 6). This data set, in total, contains 10,265 ratings as well as the corresponding recommendation information collected for three cities in the USA. The rating distributions are similar to that of Airline 3 in the previous example. We explicitly take into account the fact that the reviewing process can also lead to more extreme opinion patterns. The implications such a specific structure may have on further predictive uses of the model will be demonstrated in the next subsection. Table 6 shows the input data for the BBTRM.

At a first glance, nothing points to a specific challenge arising from this data.

Once again, the model-based estimation of the "true" ratings works properly and, as to be seen in Table 7, the resulting estimates $\hat{n}_{Y\,true}$ are very similar to their observed counterparts. The supporting outcomes of the corresponding χ^2-tests are depicted in Table 7, as well as the $\hat{\beta}$ parameters and the resulting regression parameters $\hat{a}$ and $\hat{b}$.

Obviously, in both application cases the structure of the input data can be modeled quite well. Thus, the "true" ratings can serve as a reliable starting point

Table 7 BBTRM results for the apartment data

	City 1	City 2	City 3
$\hat{\beta}_1$	0.59	0.60	0.51
$\hat{\beta}_2$	0.83	0.84	0.66
$\hat{a}$	0.11	0.11	0.10
$\hat{b}$	0.19	0.18	0.19

$$\hat{n}_{Y true} = P(r) \cdot n^{posted}$$

r	City 1	City 2	City 3
0	448	2,312	436
1	276	1,469	243
2	232	1,249	207
3	219	1,181	209
4	237	1,269	278
$\hat{n}_{true}$	1,412	7,480	1,373
H_0:	The observed frequencies $n_{Y posted}$ of the posted ratings equal their estimated counterparts $\hat{n}_{Y true}$. ($\alpha = 0.01$, $\chi^{2crit} = 9.21$; $df = 2$)		
χ^{2emp}	8.60	5.38	3.53
H_0	*Not rejected*	*Not rejected*	*Not rejected*
Conclusions	The variation of the 'true' ratings can be described by a beta distribution (H_1). The 'true' ratings can be estimated from the posted ones (H_2).		

(rounded values)

for further computations, for instance, the prediction of future sales (see, e.g., Chen et al. 2004).

Another interesting and useful application of this construct is the prediction of unknown recommendation probabilities in cases where this information is not part of the posted reviews. For example, the vendor of a product or service which is often rated on an online opinion platform might be interested in using the estimated recommendation probabilities for future demand planning or a check of the effectiveness of recent advertising efforts in the context of eWOM. This further emphasizes the importance of such additional information for both marketing research and communication management. However, information about recommendation or non-recommendation to buy is not provided by all platforms. *Amazon.com*, for example, does not survey this information. A closer consideration of this issue and the corresponding question whether the application of BBTRM might be helpful in this case will be presented in the next subsection.

3.2 Extended Use of the BBTRM Framework

In accordance with Morrison (1979), who used the "true" intentions to compute purchase probabilities, the estimated "true" ratings will now be used to determine the probability of recommending a product or service, given a certain rating r. We

Table 8 Estimated and observed recommendation frequencies for the airline data

r	$\hat{n}_{p^{rec}} = p^{rec} \cdot n_{Yposted}$				$n_{recposted}$			
	Airline 1	Airline 2	Airline 3	Airline 4	Airline 1	Airline 2	Airline 3	Airline 4
0	3	3	3	2	7	1	0	0
1	5	3	5	2	3	1	1	0
2	5	7	3	4	3	2	1	4
3	17	9	11	11	21	13	12	14
4	43	12	29	61	46	14	31	63
n	73	34	51	80	80	31	45	81

H_0: The observed recommendation frequencies $n_{recposted}$ equal their estimated counterparts $\hat{n}_{p^{rec}}$. ($\alpha = 0.01$, $\chi^{2crit} = 9.21$; $df = 2$)

χ^{2emp}					7.59	8.58	7.67	4.19
H_0					*Not rejected*	*Not rejected*	*Not rejected*	*Not rejected*

(rounded values)

therefore, once again, make use of the above two real world data sets. Since for both cases the recommendations concerning the rated objects are available, this information can be used as a benchmark for the BBTRM-based predictions of recommendation probabilities. The consideration of both examples is of particular interest because of their different accessibility for this task, regardless of the very similar performance with respect to the estimation of the "true" ratings.

The interesting recommending probability p^{rec} is defined as a function of the posted ratings. Consequently, starting from Eq. (8), we get:

$$p^{rec} = \frac{r + \hat{\beta}_1}{R + \hat{\beta}_1 + \hat{\beta}_2}. \tag{10}$$

The left side of Table 8 shows the estimated recommendation frequencies for the airline data. For comparison purpose the respective benchmark frequencies $n_{recposted}$ are displayed in the right part of the table. The value $\hat{n}_{p^{rec},Airline\ 1} = 43$ for Airline 1 then, for example, indicates that 43 of the reviewers, who assigned this airline to rating class $r = 4$, are estimated to also recommend this airline. The corresponding observation, $n_{recposted,\ Airline\ 1} = 46$, is given in the right part of the table.

All in all, the predictions are rather good, taking into account that no information about the real recommendation behavior was used in the model calibration process. The χ^2-statistics depicted at the bottom of Table 8 further support the visual impression. For all airlines, the null hypotheses of equality cannot be rejected at the 1 % level.

The above results suggest the conclusion that the BBTRM framework is a powerful tool for the prediction of unknown (missing) recommendation information. The direct computing of missing recommendation information from the "true" ratings is a significant benefit because of the fact that no other information except the ratings is required for this enhancement of the available data. However, such a good model

Table 9 Estimated and observed recommendation frequencies for the apartment data

	$\hat{n}_{p^{rec}} = p^{rec} \cdot n_{y\,posted}$			$n_{rec\,posted}$		
r	City 1	City 2	City 3	City 1	City 2	City 3
0	49	262	44	0	4	2
1	76	412	66	9	31	13
2	128	619	110	128	628	129
3	130	789	135	189	1,172	196
4	204	1,062	243	241	1,256	278
n	587	3,144	598	567	3,091	618
H_0:	The observed recommendation frequencies $n_{rec\,posted}$ equal their estimated counterparts					
	$\hat{n}_{p^{rec}}$. ($\alpha = 0.01, \chi^{2crit} = 9.21; df = 2$)					
χ^{2emp}				24,434.98	21,478.87	1,124.27
H_0				*Rejected*	*Rejected*	*Rejected*

(rounded values)

fit cannot be guaranteed in general. The corresponding results for the apartment data depicted in Table 9 show a worse model fit.

This time the predictions are rather unsatisfactory and the question is why? Obviously, the number of recommendations for the classes $r = 0$ and $r = 1$ are heavily overestimated, whereas the remaining classes feature fairly acceptable predictions. In contrast to the rather temporary service consumption process in the airline example, the permanent relevance of an apartment decision implies that negative experience results in a strict non-recommendation. Obviously, recommendations only come into question if at least a medium satisfaction (expressed by a rating equal or higher than 2) occurs. This structural split of the willingness to recommend a product or service is hard to overcome by a model which implicitly assumes a monotonous trend. For such cases another indirect approach has to be considered.

A closer look at the rating frequencies for City 1 and City 2 reveals remarkable differences regarding the total number of recommendations, as well as regarding the class-wise frequencies. The BBTRM, on the other hand, provides rather similar parameters for the linear relationship that defines the core of the model in both cases, namely $\hat{a}^{City\,1} = 0.109$ and $\hat{a}^{City\,2} = 0.110$, $\hat{b}^{City\,1} = 0.185$ and $\hat{b}^{City\,2} = 0.184$, respectively. This suggests the guess that both cases are also similar with respect to the recommendation frequencies. The data shown in Table 10 clearly supports this estimation (see the relative values).

Practically, this means that, based on a data set (e.g. the apartments offered in City 1), where recommendation frequencies are available, the unknown recommendation behavior can be predicted by transferring the recommendation pattern from one data set (City 1) to another (City 2), which offers the same parameter estimates $\hat{a}$ and $\hat{b}$ (see Table 7). In this case, for example, the recommendation behavior for City 2 could be predicted by the results for City 1, although the data of City 2 might have been unknown. Analogous approaches can be found in the literature on missing value imputation (Kalton and Kasprzyk 1982; Sande 1982). Analogies

Table 10 Absolute and relative recommendation frequencies for City 1 and City 2

	$n_{rec^{posted}}$			
r	City 1 (absolute)	City 1 (relative)	City 2 (absolute)	City 2 (relative)
0	0	0	4	0
1	9	0.02	31	0.01
2	128	0.23	628	0.20
3	189	0.33	1,172	0.38
4	241	0.43	1,256	0.41
n	567	1	3,091	1

(rounded values)

particularly exist with regard to so-called cold-deck imputation procedures. For obvious reasons a successful transfer, however, requires that both data sets represent the same product or service category.

4 Discussion and Conclusion

In recent years, online consumer review platforms have become very popular and essential for both consumers and researchers. The literature shows that two major research streams have emerged. On the one hand, researchers use the textual part of online consumer reviews for pattern mining and sentiment analysis. On the other hand, "star" ratings are used as a starting point for further analysis, for example, the prediction of future sales or purchasing behavior. The analyses regarding each research stream implicate several advantages and disadvantages. Therefore, in the following, we will first illustrate some hindrances to sentiment analysis (1). Thereafter, some advantages of the rating-based BBTRM framework are offered (2) as well as some critical aspects, which require further research (3). We finish with some general advantages of a rating-based approach (4).

1. The (automatic) analysis of review texts is a popular object of research. Nevertheless, some obvious barriers might hamper it. For example, one problem is to differentiate between subjective and objective expressions. Similar to this is the problem of automatically filtering out emotions. Furthermore, explicit and implicit features that are posted by the reviewers have to be distinguished from one another. Another problem is the differentiation between positive or negative tone, namely the determination of a meaningful sentiment value. And, on top of this, opinion mining has to handle the same general problems as other natural language processing methods: sarcasm, orthographic mistakes, common speech and so on. These aspects are widely discussed and are still a topic in linguistics (see, for example, Liu (2010) or Manning et al. (2010), who address these and some other issues in-depth).

2. In research, "star" ratings are often assumed to reflect the aggregate message of the review text regarding the satisfaction or dissatisfaction with the product or service of interest. However, there are a couple of issues which raise questions about the general reliability of ratings. Especially, overwhelmingly positive ratings give rise to suspicion (Hu et al. 2006, 2009; Li and Hitt 2010; Moe and Schweidel 2011). We elaborated on this in Sect. 1. One possibility to examine the reliability of rating data in online consumer reviews is to have a closer look at the distribution structure. The main idea behind this is that certainty about the basic distribution structure helps to ensure further analyses as, for example, the prediction of future sales. By interpreting "star" ratings as the outcome of a stochastic process, the proposed BBTRM framework can be used to investigate rating frequency distributions. In so far the estimation of the "true" ratings can be seen as a kind of internal verification of the reliability of the observed rating data. If posted ratings are replicable in the indicated sense the database can be assumed to be consistent and useful for further predictions. Considering that ratings often provide the basis for further computations with managerial implications this promises to be beneficial. The empirical usefulness of the suggested framework has been demonstrated by means of several data sets reflecting common rating frequency distributions. However, the available results also showed that a closer look at the data at hand is required before applying the framework in a given setting.

3. Structural breaks as in the case of the apartment data can significantly impair the quality of the predictions of unknown probabilities. Problems arise if the observed frequency distribution is characterized by an extreme asymmetry or if the data show a distinct multimodal structure. The question of why, for example, J-shaped data is not suitable for the BBTRM framework directly corresponds to the assumed beta distribution of the "true" ratings: In cases where the beta-distributed random variable Y^{true} can be estimated adequately, the reproduction of a broad range of empirical distributions becomes possible (see, for example, Lilien et al. 1992). Otherwise, alternative distributions are required for Y^{true}. The search and empirical verification of such distributions is work in progress and is therefore beyond the focus of this article.

Another point of discussion is the linearity assumption regarding the relationship between the posted and the estimated ratings. However, greater universality by assuming a non-linear relationship should always be contrasted with the required methodical burden going hand in hand with that. Kalwani and Silk (1982) already discussed this issue by referring to the original purchase intention framework (Morrison 1979) and suggested a piecewise linear model (see also Jamieson and Bass 1989). More flexible functions might also help to better deal with extreme rating frequency distributions and possible structural breaks in the data. Even though these challenging extensions of the basic framework have to be left to future research, the great importance of online product ratings for purchase decision making and (e)WOM, as well as for consumer satisfaction analysis and opinion mining, makes them worth a closer consideration. Last but not least one could think about an extension of the presented approach which allows the evaluation of the predicted recommendation behavior by using the full texts or the pros and cons coming along with the product ratings. But since this is not possible at the current stage of research without an expensive pre-processing of the text data we did not yet implement this idea.

4. The purely rating-based approach enables new applications in the context of international marketing and ecommerce. Particularly the independence of the full texts and, therewith, the independence of respective country-specific languages facilitates ad-hoc analyses of the present type across borders. The consideration of country-specific "true" ratings and recommendation probabilities opens new options for comparing country-specific (e)WOM behavior. From an international marketing point of view it is of increasing importance to know at which level of satisfaction (quantified by means of rating values) a population tends to recommend a product or service. In so far the new approach may also contribute to current research in international (e)WOM analysis.

Finally, the suggested framework is little sensitive to the number of observed ratings, provided basic requirements with regard to the achievement of statistically significant parameters are fulfilled. This allows the use of the BBTRM framework also in cases where only a comparatively small number of reviews are available, as in the case of the airline data above. A corresponding example is the introduction of a new product where (e)WOM already plays an important role, but consumer evaluations are still scarce.

References

Abulaish M, Jahiruddin, Doja MN, Ahmad T (2009) Feature and opinion mining for customer review summarization. In: Recognition and machine intelligence. Lecture notes in computer science, vol 5909. Springer, Berlin/Heidelberg, pp 219–224

Chen Y, Fay S, Wang Q (2003) Marketing implications of online consumer product reviews. Working paper, Department of Marketing, University of Florida

Chen PY, Wu SY, Yoon J (2004) The impact of online recommendations and consumer feedback on sales. In: Proceedings of the 24th international conference on information systems (ICIS), Washington, DC, pp 711–724

Chevalier JA, Mayzlin D (2006) The effect of word of mouth on sales: online book reviews. J Mark Res 43(3):345–354

Dellarocas C, Awad NF, Zhang XM (2004) Exploring the value of online reviews to organizations: implications for revenue forecasting and planning. In: Proceedings of the 24th international conference on information systems (ICIS), Washington, DC, pp 379–386

Dellarocas C, Gao G, Narayan R (2010) Are consumers more likely to contribute online reviews for hit or niche products? J Manag Inf Syst 27(2):127–157

Godes D, Mayzlin D (2004) Using online conversations to study word-of-mouth communication. Mark Sci 23(4):545–560

Hao L, Li X, Tan Y, Xu J (2011) The economic value of ratings in app market. http://ssrn.com/abstract=1892584

Healy MJR, Goldstein H (1978) Regression to the mean. Ann Hum Biol 5(3):277–280

Hu N, Pavlou PA, Zhang J (2006) Can online reviews reveal a product's true quality? Empirical findings and analytical modeling of online word-of-mouth communication. In: Proceedings of the 7th ACM conference on electronic commerce, Ann Arbor, pp 324–330

Hu N, Zhang J, Pavlou PA (2009) Overcoming the j-shaped distribution of product reviews. Commun ACM 52(10):144

Jamieson LF, Bass FM (1989) Adjusting stated intention measures to predict trial purchase of new products: a comparison of models and methods. J Mark Res 26(3):336–345

Jiang B, Chen P (2007) An economic analysis of online product reviews and ratings. http://ssrn.com/abstract=1023302

Joeckel S (2007) The impact of experience: the influences of user and online review ratings on the performance of video games in the us market. In: Situated play: proceedings of the 2007 digital games research association conference, Tokyo, pp 629–638. http://www.digra.org/dl/db/07312.08365.pdf

Juster FT (1966) Consumer buying intentions and purchase probability: an experiment in survey design. J Am Stat Assoc 61(315):658–696

Kalton G, Kasprzyk D (1982) Imputing for missing survey responses. In: Proceedings of the section on survey research methods, Cincinnati. American Statistical Association, pp 22–33

Kalwani MU, Silk AJ (1982) On the reliability and predictive validity of purchase intention measures. Mark Sci 1(3):243–286

Lee J, Park DH, Han I (2011) The different effects of online consumer reviews on consumers' purchase intention depending on trust in online shopping mall: an advertising perspective. Internet Res 21(2):187–206

Li X, Hitt LM (2008) Self-selection and information role of online product reviews. Inf Syst Res 19(4):456–474

Li X, Hitt LM (2010) Price effects in online product reviews: an analytical model and empirical analysis. Manag Inf Syst Q 34(4):809–831

Lilien GL, Kotler P, Moorthy KS (1992) Marketing models. Prentice Hall, Englewood Cliffs

Liu B (2010) Sentiment analysis and subjectivity. In: Indurkhya N, Damerau FJ (eds) Handbook of natural language processing. CRC, Boca Raton, pp 627–666

Liu B, Hu M, Cheng J (2005) Opinion observer: analyzing and comparing opinions on the web. In: WWW'05: proceedings of the 14th international conference on world wide web, Chiba. ACM, pp 342–351

Manning CD, Raghavan P, Schuetze H (2010) Introduction to information retrieval. Cambridge University Press, New York

Moe WW, Schweidel DA (2011) Online product opinions: incidence, evaluation and evolution. http://ssrn.com/abstract=1525205

Moe WW, Trusov M (2011) The value of social dynamics in online product ratings forums. J Mark Res 48(3):444–456

Morrison DG (1979) Purchase intentions and purchase behavior. J Mark Res 43(2):65–74

Sande IG (1982) Imputation in surveys: coping with reality. Am Stat 36(3):145–152

Thomas ML, Mullen LG, Fraedrich J (2011) Increased word-of-mouth via strategic-related marketing. Int J Nonprofit Volunt Sect Mark 16(1):36–49

Tsang ASL, Prendergast G (2009) Is a "star" worth a thousand words? The interplay between movie review texts and rating valences. Eur J Mark 43(11/12):1269–1280

Statistical Process Modelling for Machining of Inhomogeneous Mineral Subsoil

Claus Weihs, Nils Raabe, Manuel Ferreira, and Christian Rautert

Abstract Because in the machining process of concrete, tool wear and production time are very cost sensitive factors, the adaption of the tools to the particular machining processes is of major importance.

We show how statistical methods can be used to model the influences of the process parameters on the forces affecting the workpiece as well as on the chip removal rate and the wear rate of the used diamond. Based on these models a geometrical simulation model can be derived which will help to determine optimal parameter settings for specific situations.

As the machined materials are in general abrasive, usual discretized simulation methods like finite elements models can not be applied. Hence our approach is another type of discretization subdividing both material and diamond grain into Delaunay tessellations and interpreting the resulting micropart connections as predetermined breaking points. Then, the process is iteratively simulated and in each iteration the interesting entities are computed.

C. Weihs (✉) · N. Raabe
Faculty of Statistics, TU Dortmund University, Dortmund, Germany
e-mail: weihs@statistik.tu-dortmund.de; raabe@statistik.tu-dortmund.de

M. Ferreira
Institute of Materials Engineering, TU Dortmund University, Dortmund, Germany
e-mail: manuel.ferreira@tu-dortmund.de

C. Rautert
Institute of Machining Technology, TU Dortmund University, Dortmund, Germany
e-mail: rautert@isf.maschinenbau.uni-dortmund.de

W. Gaul et al. (eds.), *German-Japanese Interchange of Data Analysis Results*,
Studies in Classification, Data Analysis, and Knowledge Organization,
DOI 10.1007/978-3-319-01264-3_22,
© Springer International Publishing Switzerland 2014

1 Introduction

Tool wear and machining time represent two dominant cost factors in cutting processes. To obtain durable tools with increased performance these factors have to be optimized demanding for the investigation of the interactions between tool and workpiece. Unlike ductile materials such as steel, aluminum or plastics, material characteristics for mineral substrates like concrete are difficult to determine due to their strongly inhomogeneous components, the dispersion of the aggregates and porosities, the time dependency of the compression strength etc. (see Denkena et al. 2008). As a result of the brittleness of mineral materials and the corresponding discontinuous chip formation, there are varying engagement conditions of the tool which leads to alternating forces and spontaneous tool wear by diamond fracture.

Despite the manifold of concrete specifications, tools for concrete machining are still more or less standardized, not adapted to the particular machining application. In non-percussive cutting of mineral subsoil such as trepanning, diamond impregnated sintered tools dominate the field of machining of concrete because of the diamonds' mechanical properties. These composite materials are fabricated powder-metallurgically. Well-established techniques like cold pressing with a following vacuum sintering process or hot-pressing, being a very productive manufacturing route, are used for industrial mass production. The powder-metallurgical fabrication process implies a statistical dispersion of the diamonds embedded in the metal matrix. Additionally, the composition and allocation of different hard phases, cement and natural stone grit in the machined concrete are randomly distributed. Because of these facts, the exact knowledge of the machining process is necessary to be able to investigate for appropriate tool design and development.

To obtain a better understanding of these highly complex grinding mechanisms of inhomogeneous materials which can not be described solely by physical means, statistical methods are used to take into account the effect of diamond grain orientation, the disposition of diamonds in the metal matrix and the stochastic nature of the machining processes of brittle materials. The first step to gain more information about the machining process is the realization of single grain wear tests on different natural stone slabs and cement.

2 Experimental Setup

To gain information about the fundamental correlations between process parameters and workpiece specifications, single grain scratch tests have been accomplished. For these, isolated diamond grains, brazed on steel pikes have been manufactured (see Fig. 1, left and middle) to prevent side effects from the binder phase or preceded grain scratches as they occur in the grinding segments in real life application. To provide consistent workpiece properties, high strength concrete specimens of specification DIN 1045-1, C80/90 containing basalt as the only aggregate had been

Fig. 1 Sample before and after brazing (*left* and *middle*) and scratch test device on basalt (*right*)

produced. Also, the two phases, cement binder and basalt, were separately prepared as homogeneous specimens for an analysis of the material specific influence on the wear.

To eliminate further side effects such as hydrodynamic lubrication and interaction of previously removed material and adhesion, the experiments have been carried out without any coolant. The brazed grain pikes had been attached to a rotating disc which in turn had been mounted to the machine (see Fig. 1, right) to simulate the original process kinematic. Parameters for the experimental design were chosen according to common tools and trepanning processes. To guarantee constant depth of cut the rotatory motion of the grain pike had been superimposed by a constant feed which generated a helical trajectory. To generate a measurable grain wear, a certain distance had to be accomplished. A total depth of cut of 250 μm had been achieved in every test.

3 Design of Experiments and Regression Models

In order to investigate the influences of the process parameters to the responses tool wear and forces, a series of 92 single grain scratch experiments had been carried out based on a statistical design of experiments. Table 1 shows the factor levels of this design. Note that because the outer hole diameter d_p (in *mm*) is not adjustable continuously a twice-replicated Central Composite Design (CCD, Eriksson 2008) of the cutting speed v_p (in *rpm*) and the depth of cut a_p (in $\mu m/r$) with seven center point runs had been repeated for each of the four diameter levels.

In a first attempt to provide a basis for the simulation models the results of these experiments had been analyzed by fitting regression models using stepwise forward-backward-selection based on the Akaike Information Criterion taking into account linear, quadratic and two-fold interaction effects. This procedure led to the following equations for average normal and radial forces $\bar{F}_n$ and $\bar{F}_r$ (in N) and for the tool wear measured by the height decrease Δh of the diamond grain:

$$\hat{\bar{F}}_n = 44.025 - 0.524d_p - 0.351a_p^2 + 0.062d_pa_p,$$

Table 1 Factor levels of the CCD

Factor	Levels				
Outer hole diameter d_p	50	80	110	130	
Depth of cut a_p	3.75	5	7.5	10	11.25
Cutting speed v_p	346	525	900	1,275	1,454

$$\hat{\bar{F}}_r = -5.204 + 0.007v_p + 2.425a_p - 0.187a_p^2 - 0.00009v_pd_p + 0.007d_pa_p,$$

$$log(\hat{\Delta}h) = -10.67 + 1.386a_p - 0.069a_p^2 + 0.00007d_p^2.$$

All model coefficients are significant on a level of 5 % and the R^2s are $0.181, 0.236$, and 0.131. The low goodness of fit has to be seen in the context of the relatively high reproduction variance (i.e. the variance of realizations with the same factor levels), due to which the R^2 values are limited by $0.652, 0.628$, and 0.395.

4 Simulation Model

The regression formulas of the previous section had been used to fit the parameters of a geometrical physical simulation model of tool and workpiece. The main idea behind this model is to see the abrasive materials as bonds of microparts which are flattened and broken out during the process. Our approach is to determine these bonds by the Delaunay tessellations of sets of points which are uniformly distributed within the corresponding workpiece or grain shape (see Fig. 2). Thereby the workpiece is reduced to a ring covering the scratch line to be produced and the grain is idealized to the shape that is formed when the corners of a cube are flattened, where the amount of the reduction due to flattening is described by two parameters k_k and v_k (see Fig. 2).

Positions of Workpiece and Diamond: For the simulation of the process the diamond grain and the workpiece are adjusted according to the settings of the corresponding machining process leading to the initial coordinates of workpiece and grain vertices S_w and S_k. The position of the workpiece stays constant during the whole process and is centered around the origin parallel to the xy-plane, where the z-coordinate of the bottom is $z = 0$. The grain is first centered in the origin and successively turned around the (x, z, y)-axes by angels α_x, α_z and α_y. Finally, the grain is moved to its starting position by shifting all grain vertices along the x-axis by the drilling diameter d_p and along the z-axis by the starting height $h_{k;0}$.

Beside d_p in mm the depth of cut a_p in $\mu m/r$ and the cutting speed v_p in rpm are the process parameters. With sampling rate r_p in Hz, the actual position of the grain at the beginning of each iteration i can be determined by turning the grain vertices around the z-axis and shifting them along the same axis:

Fig. 2 Edge length
reductions and Delaunay
tessellations of randomly
distributed points

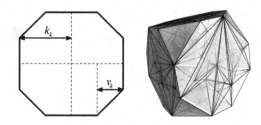

$$S_k^i = S_k^{i-1} \begin{pmatrix} cos\alpha_r & 0 & sin\alpha_r \\ 0 & 1 & 0 \\ -sin\alpha_r & 0 & cos\alpha_r \end{pmatrix} - \begin{pmatrix} 0 & a_r & 0 \\ \vdots & \vdots & \vdots \\ 0 & a_r & 0 \end{pmatrix}$$

with $\alpha_r = (2\pi v_p)/(60 r_p)$, $a_r = (a_p v_p)/(60000 r_p)$.

Workpiece Affecting Forces: Next, in each iteration the matrix W_s of intersection
volumes of all pairs of grain and workpiece simplices is computed, where the
(l, j)th entry $w_{s;lj}$ of W_s reflects the intersection volume of the lth grain simplex
with the jth workpiece simplex. Then, for each workpiece simplex intersecting at
least one grain simplex it is determined, which force affects it. For this purpose
first the total mass $m_{k;j} = \sum_{l:w_{s;lj}>0} w_{k;l} \rho_k$ of all grain simplices that intersect the
actual workpiece simplex j is computed from the grain simplex volumes $w_{k;l}$ and
the grain density ρ_k. By interpreting the collision of workpiece and grain simplices
as a force impact the force affecting the jth workpiece simplex can be simulated
by assuming the relation $F_{ij} = (v_p m_{k;j})/t_d$, where the constant t_d is the iteration
length.

How the workpiece affecting force F_{ij} distributes in radial and normal direction
depends on the geometrical properties of the involved simplices. In this context
for simplicity only the largest grain simplex is considered and the vertices of this
simplex as well as those of the workpiece simplex are projected to the vertical plane
parallel to the cutting direction. In this projection the angle γ_k between the vector
orthogonal to the cutting direction and that vertex of the grain simplex, that contacts
the workpiece simplex first, is computed. After determining the corresponding angle
γ_w the normal and radial forces can be computed by $F_{n;ij} = F_{ij} sin[max(\gamma_w, \gamma_k)]$
and $F_{r;ij} = F_{ij} cos[max(\gamma_w, \gamma_k)]$ (see Fig. 3).

The total forces affecting the workpiece in iteration i are then given by

$$F_{n;i} = \sum_{j:\sum_{l=1}^{n_{s;k}} w_{s;lj}>0} F_{n;ij} \text{ and } F_{r;i} = \sum_{j:\sum_{l=1}^{n_{s;k}} w_{s;lj}>0} F_{r;ij}.$$

The material removal rate of the workpiece and the wear rate of the grain at the
end of each iteration is then given by the change of the corresponding volumes.

Fig. 3 Geometric
distribution of forces on
collision of grain and
workpiece simplex

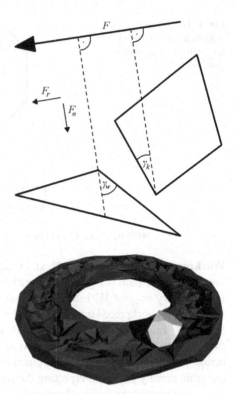

Fig. 4 Simulated machined
workpiece and diamond grain

Figure 4 shows an exemplary view of a machined workpiece and a diamond
grain, for better visualization in unrealistic proportions.

Tool Wear: The simplices of the workpiece always break out when hit by a grain
and the continuous wear of the grain is modeled by decreasing all simplices in
action by a specific factor η_k. However, when the mass $w_{w;j}$ of the active workpiece
simplex exceeds a threshold defined by the inequality $w_{w;j}\rho_w > \mu_k m_{k;j}$ with a
specific constant μ_k, the density ρ_w of workpiece material and mass $m_{k;j}$ of active
grain simplex, the grain simplex also breaks out. The material removal rate of the
workpiece and the wear rate of the grain at the end of each iteration is then given by
the change of the corresponding volumes.

Figure 5 shows the pseudo-code of the simulation model.

Model Fitting: For the optimal determination of the unknown simulation model
parameters μ_k and η_k and of the iteration length t_d a two step procedure is applied
(with the other model parameters fixed, e.g. as $r_p = 10,000\,\text{Hz}, \rho_w = 2, \rho_k = 3.52, k_k = 1, v_k = 0.7, \alpha_x = \alpha_z = \alpha_y = \pi/4$). See Raabe et al. (2011) for
more details on the simulation model. In the first step parameters are chosen by
minimizing the deviation of simulated and real response data based on a statistical
design and in the second step the model is calibrated by introducing some process

compute (S_k, S_w)
$S_k^0 \leftarrow S_k R_x R_z R_y + (d_p, h_{k;0}, 0) \otimes 1_{n_k}$, $n_k = $ no. of vertices
for $i = 1 \rightarrow i_{max}$ **do**
    $S_k^i \leftarrow S_k^{i-1} R_r - (0, a_r, 0) \otimes 1_{n_k}$
    compute intersection volumes W_s
    for $j = 1 \rightarrow n_w$ **do**
        $m_{k;j} \leftarrow \sum_{l:w_{s;lj}>0} w_{k;l} \rho_k$
        compute (γ_w, γ_k)
        $\gamma \leftarrow max(\gamma_w, \gamma_k)$
        $F_{ij} \leftarrow (v_p m_{k;j})/t_d$
        $(F_{n;ij}, F_{r;ij}) \leftarrow F_{ij}(sin\gamma, cos\gamma)$
        if $w_{w;j} \rho_w > \mu_k m_{k;j}$ **then**
            remove diamond simplices $l : w_{s;lj} > 0$
        else
            reduce heights of diamond simplices $l : w_{s;lj} > 0$ by η_k
        end if
        remove workpiece simplex j
    end for
    $(F_{n;i}, F_{r;i}) \leftarrow (\sum_j F_{n;ij}, \sum_j F_{r;ij})$
end for

Fig. 5 Pseudocode representation of simulation model

Table 2 Factor levels of model parameters

Factor	Levels		
Specific constant μ_k	5	15	25
Specific factor η_k	0.01	0.025	0.04
Reciprocal sample rate t_d	10	55	100

parameter dependent scale terms adjusting for the remaining discrepancy of the equations in Sect. 3 and the corresponding regressions applied to the simulated data.

In the first step, based on a combined design with $8 \times 27 = 216$ experiments consisting of the process parameter settings of the cube of the CCD for the two middle diameter levels (see Table 1) replicated for each setting of a full factorial 3^3-Design (with levels in Table 2) for the model parameters a quadratic model of the squared deviations of simulated and measured forces is fitted in dependence of the model parameters. By minimizing the squared Euclidian distance between the simulated and realized two-dimensional force vectors the model parameters are estimated by $\mu_k = 10.638, \eta_k = 0.036$, and $t_d = 39.8$.

In the second step, the model calibration, process parameter dependent scale terms $s_r(v_p, a_p, d_p)$ and $s_n(v_p, a_p, d_p)$ for the simulated forces are introduced to adjust for the remaining systematic model deviations. These scale terms are obtained by fitting the models $F_{r;M}/F_{r;S} = (v_p \ a_p \ d_p)\beta_r + \varepsilon_r$ and $F_{n;M}/F_{n;S} = (v_p \ a_p \ d_p)\beta_n + \varepsilon_n$. The estimated ratios are taken as the scale terms and are stored as multiplicative factors for the simulated forces.

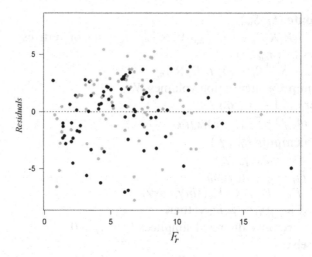

Fig. 6 Comparison of cross residuals for F_r based on measured (*grey*) and simulated (*black*) data (Color figure online)

With the simulation model calibrated this way, the whole statistical design had been repeated and by comparing the results of this simulation series with the real experiments it can be shown that simulated and real data are very close. This impression is exemplarily emphasized by a comparison of the cross residuals, i.e. the residuals from applying the models for simulated forces to measured forces and vice versa (Fig. 6). Moreover, in order to decide, whether the differences between the models are significant, joint models for simulated and measured forces are fitted. In these models, an indicator variable I_F showing if a specific data point had been measured ($I_F = -1$) or simulated ($I_F = +1$) is introduced both as main effect as well as in interactions with all contained regressors. Since for both forces I_F had no significant influence in the models, neither as main effect nor in any interaction (see Table 3), the results confirm that the same models hold for both simulated and measured forces.

4.1 Extensions

As shown in the previous section the simulation model is able to reflect the process behavior quite well. However, it still neglects some important factors, namely the wear of the grain and the heterogeneity of the material which causes highly non-stationary processes.

For the investigation of the first factor the influence of the grains had been studied by including them into the regression analysis. There, the multi-used grains are directly taken into account as random effects u (included in the model by right

Table 3 Results of the joint regression models of simulated and measured force data. Coefficients significant on a level of 5 % bold

Regressor	Model for F_n			Model for F_r		
	Coefficient	p-value		Coefficient	p-value	
Intercept	**42.666**	**<0.0001**		**−9.184**	**0.0031**	
v_p				**0.0055**	**0.0079**	
a_p				**3.671**	**<0.0001**	
d_p	**−0.4934**	**0.0003**				
a_p^2	**−0.3399**	**0.0018**		**−0.253**	**<0.0001**	
$v_p d_p$				**−0.00006**	**0.01**	
$a_p d_p$	**0.0599**	**0.0005**		0.00406	0.106	
I_F	−1.0435	0.8868		−2.956	0.3358	
$I_F v_p$				−0.00254	0.2277	
$I_F a_p$				1.048	0.2143	
$I_F d_p$	0.02635	0.8419				
$I_F a_p^2$	0.0027	0.9801		−0.0587	0.3413	
$I_F v_p d_p$				0.000035	0.102	
$I_F a_p d_p$	−0.00048	0.9774		−0.00031	0.2114	

Table 4 Regression models of forces and tool wear

Target	Model	R_{loo}^2
$\bar{F}_n$	$= 44.025 - 0.524 d_p - 0.351 a_p^2 + 0.062 d_p a_p + \varepsilon, \varepsilon \sim N(0, 9.458)$	0.118
	$= 16.697 + 0.025 v_p - 0.321 d_p - 0.003 v_p a_p + 0.037 a_p d_p + 9.968 x$ $+ Du + \varepsilon, u \sim N(0, 8.604), \quad \varepsilon \sim N(0, 6.66)$	0.474
$\bar{F}_r$	$= -5.204 + 0.007 v_p + 2.425 a_p - 0.187 a_p^2 - 0.00009 v_p d_p + 0.007 d_p a_p$ $+ \varepsilon \varepsilon \sim N(0, 2.976)$	0.114
	$= 1.408 + 0.009 v_p - 0.039 a_p^2 - 0.0001 v_p d_p + 0.008 a_p d_p + 2.35 x + Du$ $+ \varepsilon, \quad u \sim N(0, 2.36), \quad \varepsilon \sim N(0, 2.35)$	0.300
$log(\Delta h)$	$= -10.67 + 1.386 a_p - 0.069 a_p^2 + 0.00007 d_p^2 + \varepsilon, \varepsilon \sim N(0, 1.644)$	0.128
	$= -12.88 + 1.85 a_p + 0.000001 v_p^2 - 0.057 a_p^2 + 0.0003 d_p^2 - 0.007 a_p d_p$ $-1.02 y + Du + \varepsilon, \quad u \sim N(0, 0.914), \quad \varepsilon \sim N(0, 1.329)$	0.24

multiplication to the random effect design matrix D). Additionally the orientation of the grains is analyzed (see Tillmann et al. 2011) leading to a pair of coordinates x and y, which are also included in the regression analysis.

By these additions the low leave-one-out cross validated multiple correlation coefficients R_{loo}^2 of 0.118 ($\bar{F}_n$), 0.114 ($\bar{F}_r$) and 0.128 (Δh) for the three regressions could be improved significantly. By repeating the stepwise model selection with the orientation and grain data the R_{loo}^2 values increase to the values shown in Table 4, which also summarizes the corresponding models for each of the target variables.

The second important factor neglected in the actual simulation model is the heterogeneity of the material. This factor will become even more severe when more heterogeneous materials like armored concrete will be studied. The consideration of the heterogeneity demands for the investigation of the force time series. A closer

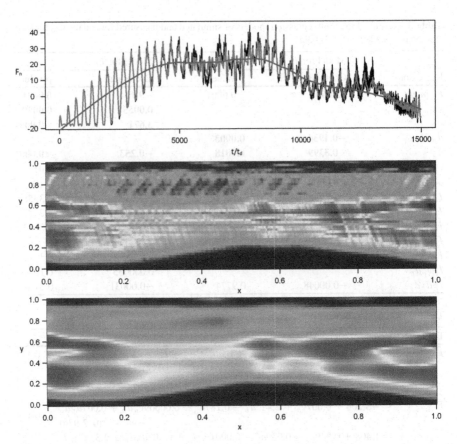

Fig. 7 *Top*: exemplary process with time smoothed (*red*) and spatially smoothed (*green*) paths. *Middle*: force image of drilling wall. *Bottom*: spatially smoothed force image of drilling wall (Color figure online)

look to these reveals the high non-stationarity of the process. As the scratch lines are concentric circles with fixed radius and the rotational frequency of the tool is also fixed, it is possible to construct images of the forces by mapping the sample points to the corresponding positions at the drilled wall for each process. These images are typically very noisy due to the high variability in measuring the forces and therefore have to be smoothed spatially by an appropriate procedure. Figure 7 shows a typical process signal and corresponding force images where the lower image had been smoothed using local polynomial regression in the x- and y-coordinates.

We plan to include the distributions and equations derived from the investigations of grain orientation and material heterogeneity into the simulation model to make it more realistic and to be able to test optimization strategies for the machining process in a more reliable way.

Acknowledgements We thank the German Science Foundation (DFG) for its support in the Collaborative Research Center (SFB) 823.

References

Denkena B, Boehnke D, Konopatzki B, Buhl JC, Rahman S, Robben L (2008) Sonic analysis in cut-off grinding of concrete. Prod Eng 2(2):209–218

Eriksson L (2008) Design of experiments: principles and applications. MKS Umetrics AB, Malmö

Raabe N, Rautert C, Ferreira M, Weihs C (2011) Geometrical process modeling of concrete machining based on delaunay tessellations. In: World congress on engineering and computer sciences 2011, San Francisco, vol 2, pp 991–996

Tillmann W, Biermann D, Ferreira M, Rautert C, Raabe N, Trautmann H (2011) Crystal orientation determination of monocrystalline diamond grains for tool application with electron backscatter diffraction (EBSD). In: Euro PM2011 Proceedings, Barcelona, vol 1, pp 265–270

Index

W. Gaul et al. (eds.), *German-Japanese Interchange of Data Analysis Results*,
Studies in Classification, Data Analysis, and Knowledge Organization,
DOI 10.1007/978-3-319-01264-3,
© Springer International Publishing Switzerland 2014